Exploring The BUILDING BLOCKS of Science

Book 6

LABORATORY NOTEBOOK

ASTRONOMY

Geology

CHEMISTRY

Biology

Physics

REAL SCIENCE 4 Kids

REBECCA W. KELLER, PhD

Illustrations: Janet Moneymaker

Exploring the Building Blocks of Science Book 6 Laboratory Notebook
ISBN 978-1-941181-14-0

Published by Gravitas Publications Inc.
Real Science-4-Kids®
www.realscience4kids.com
www.gravitaspublications.com

GRAVITAS
PUBLICATIONS

Contents

CHEMISTRY

BIOLOGY

PHYSICS

GEOLOGY

ASTRONOMY

Experiment 1

Take It Apart!

A printed circuit board and components
Photo credit: Warren Gretz/NREL

Introduction

Look inside an electronic device to see what's there.

I. Think About It

❶ What kinds of materials do you think a camera is made of? Plastic? Metal? Other? How is it put together?

_____

_____

_____

_____

❷ What kinds of materials do you think a cell phone is made of? Plastic? Metal? Other? How is it put together?

_____

_____

_____

_____

❸ What kinds of materials do you think a radio is made of? Plastic? Metal? Other? How is it put together?

_____

_____

_____

_____

❹ What kinds of materials do you think a television is made of? Plastic? Metal? Other? How is it put together?

_____

_____

_____

_____

_____

❺ What kinds of materials do you think a computer is made of? Plastic? Metal? Other? How is it put together?

_____

_____

_____

_____

_____

❻ How are a camera, cell phone, radio, television set, and computer similar? How are they different?

_____

_____

_____

_____

_____

II. Experiment 1: Take It Apart!   Date _____

Objective _____

_____

Hypothesis _____

_____

_____

Materials

an old digital camera, cell phone, radio, or other small electronic device
small tools such as screwdriver, tweezers, pick
rubber gloves
library or internet resources

EXPERIMENT

❶ Find a piece of technology that you have used recently such as a cell phone, camera, or radio that is no longer needed and that you have permission to disassemble.

❷ Observe the outside of the item and list below all of the materials that the item is composed of—for example, plastic, metal (what kind?), glass, etc.

_____

_____

_____

_____

_____

_____

_____

❸ Using the appropriate tools and wearing rubber gloves, carefully disassemble the item. Take off the back or outer covering and observe the inside. Take out any parts you can reasonably disassemble. Do not try to open batteries. In the space below, draw the parts and list what they are made of.

Parts of a _____

Results

❶ Using the library or an internet resource, research how the item was made, where it was made, what it is made of, and who designed it.

Facts about _____

❷ Using the library or an internet resource, explore the scientific processes that were involved in making the item and explain in what ways chemistry, physics, biology, geology, and/or astronomy were involved in the design, craft, and assembly of the item.

Sciences and processes involved in making _____

III. Conclusions

Based on your observations, what conclusions can you draw from the results of your investigation?

_____

_____

_____

_____

_____

_____

_____

_____

_____

_____

_____

_____

_____

_____

_____

_____

_____

IV. Why?

As we use a cell phone or snap a photo with a camera, we seldom think about the science behind the device we are using. But all technological devices, including the instruments scientists use, were developed as a result of scientific discoveries and technological invention. For example, all modern electronic devices contain some type of PCB (printed circuit board) that is created by a process called *chemical etching*. Without understanding how materials react chemically, PCBs would not be possible.

Printed circuit boards are made up of layers of silicon that are pressed together to form a *wafer*. A thin layer of copper is then pressed on top of the wafer. Copper is an excellent conductor of electricity, making it a useful metal for circuit boards and other electronic devices. The copper on the PCB is formed into thin lines that act like tiny wires to conduct electricity. To make the thin copper lines, acids are used to remove the unwanted copper between the lines. This is the chemical etching process.

During the etching process, the pattern of thin lines of copper that is to be kept is covered with a mask made of a substance that doesn't react with the acids. The copper beneath the masked off part is protected from the acid. Engineers designing new circuits use computer programs to design the masks for PCBs. The mask is usually printed right onto the circuit board. (This is why they are called "printed" circuit boards.) The masking material covers the narrow lines of copper and leaves the unwanted areas of copper exposed.

Once the mask is in place, the etching agent used to remove the exposed copper is applied to the masked surface. In factories where PCBs are mass-produced, the etching agent might be sprayed onto the surface or the wafer might be dipped into the acid. The etching agent reacts with the exposed copper, which then dissolves.

After the etching process is complete, the surface of the wafer must be rinsed to remove the etching agent. Water is often used for rinsing because it will dilute the acid and wash it away. Once the etching agent has been thoroughly rinsed off, the mask must then be removed. Chemicals like acetone or alcohol are often used to remove the masking material.

The thin copper lines left on the surface of the PCB are called *traces*. The traces will be in the exact pattern of the mask that was applied. Different electronic components can be attached to the PCB, and the traces will act like tiny wires that carry electricity between the components. The completed PCB along with its components will be installed in an electronic device.

V. Just For Fun

Use a glass etching kit to explore how chemical etching works on glass. What kinds of patterns can you make? Can you etch drawings into glass? What might you create by knowing how to use chemical etching? In the following spaces, write and draw your ideas and results.

Ideas & Results for Chemical Etching of Glass

Ideas & Results for Chemical Etching of Glass

Experiment 2

Reading the Meniscus

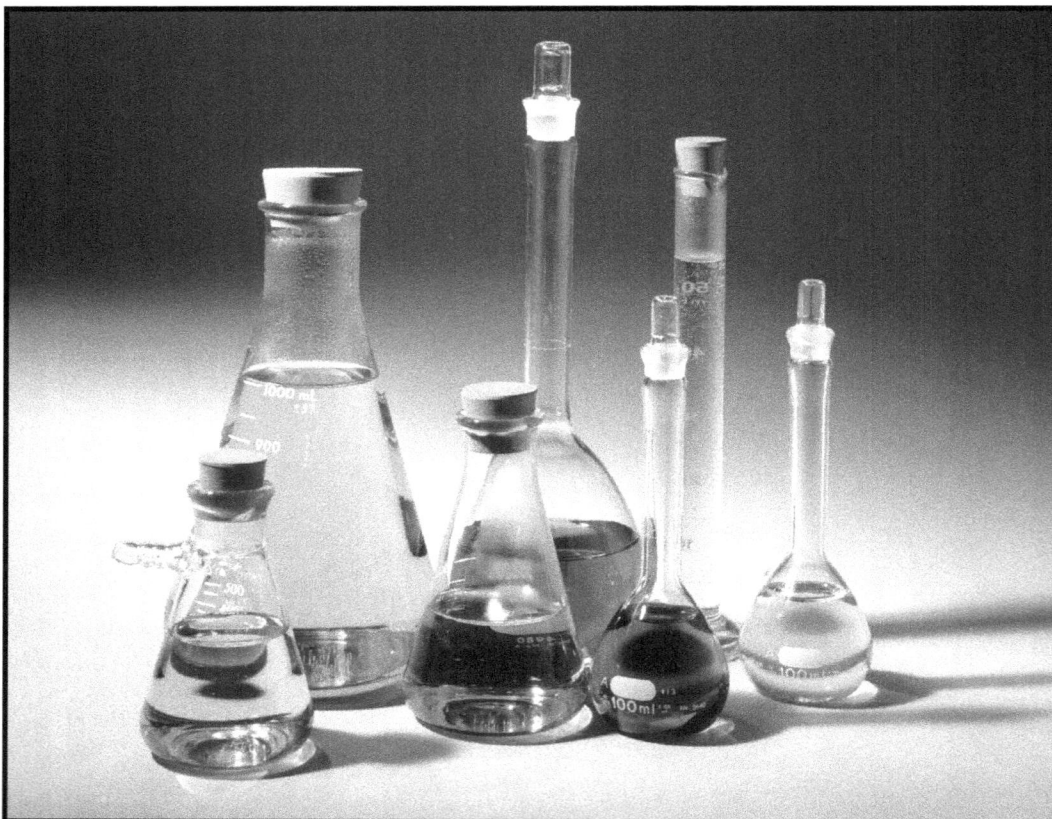

Photo Credit: Warren Gretz/National Renewable Energy Laboratory (NREL)

Introduction

How are measurements affected when water or oil meets glass?

I. Think About It

❶ If you put water in a glass cup or jar, do you think the surface of the water will be completely level? Why or why not?

_____

_____

_____

_____

❷ If you put water in a glass cup or jar, what do you think will happen to the water that is against the glass? Why?

_____

_____

_____

_____

❸ Do you think water will act the same way when it meets plastic as it does when it meets glass? Why or why not?

_____

_____

_____

❹ Do you think when oil meets glass it acts in the same way as water does when it meets glass? Why or why not?

_____

_____

_____

_____

_____

❺ Have you observed a mercury thermometer? What does the mercury look like where it is against the sides of the glass thermometer? Why?

_____

_____

_____

_____

_____

❻ Do you think it would be easy or difficult to measure water or oil accurately? Why?

_____

_____

_____

_____

_____

II. Experiment 2: Reading the Meniscus   Date _____

Objective _____

_____

Hypothesis _____

_____

_____

CHEMISTRY

Materials

    10 ml glass graduated cylinder
    glass eyedropper
    60 ml (1/4 cup) water
    60 ml (1/4 cup) rubbing alcohol
    60 ml (1/4 cup) vegetable oil

EXPERIMENT

❶ Study the graduated cylinder for a few minutes. Notice the details—the width of the mouth, the pour spout, the markings along the side. You should see markings that correspond to volumes in 1 ml increments.

❷ Pour water into the graduated cylinder until the water reaches the 5 ml mark.

❸ Hold the graduated cylinder so the top of the water in the cylinder is at eye level. Look at the sides of the graduated cylinder and observe whether the water is clinging to the sides, creating a curve. What else can you notice? In the *Results* section record your observations.

❹ The curved line of the water is called the *meniscus*. Look for the bottom of the meniscus (lowest point of the curvature). Is the bottom above or below the 5 ml mark?

CHEMISTRY

❺   If the bottom is below the 5 ml mark, add a few drops of water with the eyedropper until the bottom of the meniscus is exactly at the 5 ml mark. If the bottom is above the 5 ml mark, carefully pour a little water out until the bottom of the meniscus is exactly at the 5 ml mark. You may have to pour water out and add water several times before you can get the bottom of the meniscus to match the 5 ml mark. In the *Results* section record how easy or difficult it is to get the bottom of the meniscus to match the 5 ml mark.

❻   Pour out the water and repeat Steps ❷-❺ with rubbing alcohol.

❼   Pour out the rubbing alcohol and repeat Steps ❷-❺ with vegetable oil.

Results	Observations of the meniscus of the liquid in the graduated cylinder	Notes about the ease or difficulty of measuring the liquid to exactly 5 ml
Water		
Rubbing Alcohol		
Vegetable Oil		

III. Conclusions

What conclusions can you draw from the results of your investigation?

_____

_____

_____

_____

_____

_____

_____

_____

_____

_____

_____

_____

_____

_____

_____

_____

_____

CHEMISTRY

CHEMISTRY

IV. Why?

Materials such as glass can attract or repel different liquids. This causes a liquid in glassware to form a curve, or meniscus, at its upper surface making it difficult to get an accurate reading in a measuring device such a graduated cylinder.

Water is attracted to glass. When you pour water into a graduated cylinder made of glass, the water likes the glass so much it will try to spread out on the sides of the graduated cylinder. As a result, the water will crawl up the glass, creating a concave meniscus. In a concave meniscus the water in contact with the glass is higher than the water in the center, or farther from the glass. To get an accurate reading with a concave meniscus, it is important to align the bottom of the meniscus to the graduated marks on the cylinder rather than reading from the highest part of the water where it is in contact with the glass.

When you pour oil into a glass graduated cylinder, the oil is repelled by the glass. In an effort to avoid the glass sides, the oil will pull downward, creating a convex meniscus. To get an accurate reading with a convex meniscus, it is important to align the top of the meniscus to the graduated marks on the cylinder rather than taking a reading at the point where the oil is in contact with the glass.

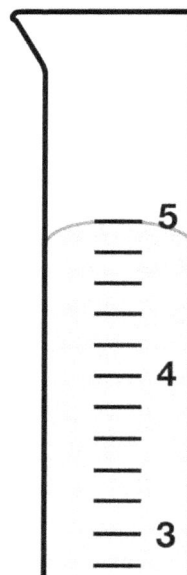

V. Just For Fun

Can you change the way water and oil form a meniscus?

❶ Coat the inside of a glass graduated cylinder or a disposable glass tube with a water repellent material such as liquid car wax, floor wax, silicone spray, or Scotch-Gard. Pour water into the graduated cylinder. Observe the meniscus and record your observations.

❷ Empty the graduated cylinder and then pour oil into it. Observe the meniscus and record your observations.

	Observations of the meniscus of the liquid in the waxed glass graduated cylinder	Comparison of the meniscuses in the two experiments
Water		
Vegetable Oil		

Experiment 3

Making an Acid-Base Indicator

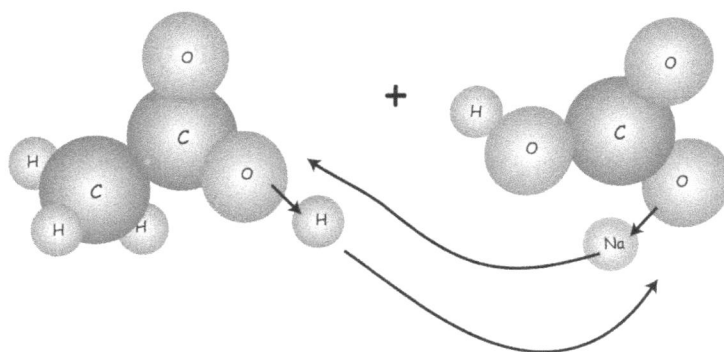

Introduction

Make your own acid-base indicator and use it to test different solutions.

I. Think About It

❶ What solutions do you think are acids?

_____

_____

_____

_____

_____

❷ What solutions do you think are bases?

_____

_____

_____

_____

_____

❸ Do you think it can be useful to know if a solution is an acid or a base? Why or why not?

_____

_____

_____

_____

_____

❹   How would find out if a solution is an acid or a base?

_____

_____

_____

_____

❺   Do you think acids can be useful? Harmful? Why or why not?

_____

_____

_____

_____

_____

❻   Do you think scientists might want to know whether two solutions are acids or bases before they mix them together? Why or why not?

_____

_____

_____

_____

II. Experiment 3: Making an Acid-Base Indicator

Date _____

Objective _____

_____

Hypothesis _____

_____

_____

Materials

one head of red cabbage
distilled water
various solutions, such as:
    ammonia
    vinegar
    clear soda pop
    milk
    mineral water
large saucepan

knife
several small jars
white coffee filters
eyedropper
measuring cup
measuring spoons
marking pen
scissors
ruler

EXPERIMENT

❶ Take the whole head of red cabbage and divide it into several pieces.

❷ Place about .7 liter (3 cups) of distilled water in a large saucepan and bring the water to a boil. Place the cabbage in the boiling water and boil for several minutes.

❸ Remove the cabbage and let the water cool. The water should be a deep purple color.

❹ Take .25 liter (1 cup) of the cabbage water to use in this experiment, and REFRIGERATE the rest for the next experiment.

CHEMISTRY

❺   Cut the coffee filters into small strips about 2 cm (3/4 inch) wide and 4 cm (1 1/2 inches) long. Make at least 20.

❻   Using the eyedropper, put several drops of the cabbage water onto each of the filter paper pieces and allow them to dry. They should be slightly pink and uniform in color. If the papers are too light, more solution can be dropped onto them, and they can be dried again. These are your acid-base indicator (pH) papers.

❼   Label one of the jars **Control Acid**, and place 15 ml (1 tbsp.) of vinegar in the jar. Add 75 ml (5 tbsp.) of distilled water. This is your known acid.

Label another jar **Control Base** and add 15 ml (1 tbsp.) of ammonia to the jar. Add 75 ml (5 tbsp.) of distilled water. This is your known base.

Label a separate jar for each of the solutions you will be testing. Put into the appropriate jar 15 ml (1 tbsp.) of each of the solutions you have collected, and add 30-75 ml (2-5 tbsp.) of distilled water to each jar.

❽   Carefully dip a strip of pH paper into the **Control Acid**. Look immediately at the pH paper for a color change and record your results in the chart on the next page. Then tape the pH paper in the pH Paper Sample column next to the section labeled **Control Acid**.

❾   Carefully dip an unused piece of pH paper into the **Control Base**. Look immediately at the pH paper for a color change, and record your results in the chart on the next page. Tape the pH paper in the space next to the **Control Base** section.

❿   Now take unused pieces of pH paper, and dip them into the other solutions you have made. Record your results. Tape the papers into the chart.

Results

pH Paper Sample	Name of Solution	Color of pH Paper	Acid or Base?
	Control Base:		
	Control Acid:		

III. Conclusions

What conclusions can you draw from your observations?

_____

_____

_____

_____

_____

_____

_____

_____

_____

_____

_____

_____

_____

_____

_____

CHEMISTRY

IV. Why?

In this experiment you made your own pH paper using red cabbage juice. Before pH meters were invented, pH paper was the most common way to test for acids and bases. pH paper can still be found in most laboratories. Before pH paper and other modern techniques were available, many chemists tasted things to find out more about them. However, this is quite dangerous, and today scientists do not taste anything in the laboratory.

The exact pH of the solutions you tested couldn't be determined by the pH paper you made. You could only find out whether a solution was acidic or basic. pH paper is made with a compound called an *indicator*. An indicator is any molecule that changes color as a result of a pH change. The molecules that give red cabbage its color react differently with acids and bases, turning pink in the presence of acids and green in the presence of bases.

The properties of acids and bases are quite different, and in many ways opposite. Acids are sour, not slippery, and are effective in dissolving metals. Bases are bitter, slippery, and react with metals to form precipitates. Because some acids and bases can be harmful, scientists do not test unknown solutions by putting them on their skin.

In this experiment you used *controls*. A control is a part of an experiment where the outcome is already known or where a given outcome can be determined. The control provides a point of reference or comparison for an experiment that uses unknowns. For example, in this experiment you tested for acidity or basicity with a pH indicator, but you did not know what the expected color change would be. By using solutions that are known to be either acidic (vinegar) or basic (ammonia), you could determine what the color change for an acid would be and what the color change for a base would be. Only then could you determine the meaning of the results of testing the unknowns.

A control can also tell the scientist when an experiment has failed. If a color change is observed in the control experiment but not in the new experiment, something may be wrong with the setup or design of the new experiment. Control experiments help scientists check for errors.

CHEMISTRY

V. Just For Fun

See whether different natural materials can be used as acid–base indicators.

Crush the material you will be testing and put some of it in each of two small jars. Using the solutions you have identified as acid or base, add some acid to one jar and some base to the other jar. Does the color change? Record your results in the chart that follows.

Natural Materials for Experimentation

Turmeric
Poppyseed or cornflower petals
Madder plant (Rubiaceae family)
Red beets
Rose petals
Berries
Blue and red grapes
Cherries
Geranium petals
Morning glory
Red onion
Petunia petals
Hibiscus petals (or hibiscus tea)
Carrots
Other natural materials of your choice

Natural Material and Its Color	Name of Solution Used	Acid or Base?	Final Color/ Indicator?

Experiment 4

Vinegar and Ammonia in the Balance:
An Introduction to Titration

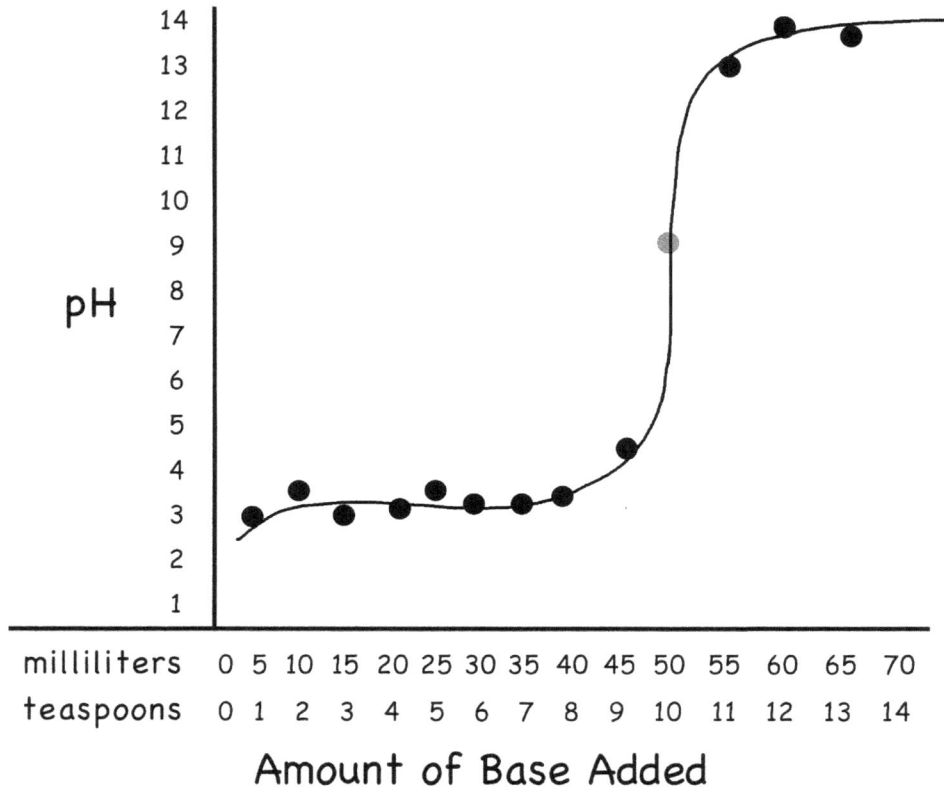

pH vs. Amount of Base Added

| milliliters | 0 5 10 15 20 25 30 35 40 45 50 55 60 65 70 |
| teaspoons | 0 1 2 3 4 5 6 7 8 9 10 11 12 13 14 |

Amount of Base Added

Introduction

Do you think you can tell when an acid is neutralized by a base?

I. Think About It

❶ Do you think it can be important to know the concentration of an acid or a base? Why or why not?

_____

_____

_____

_____

❷ What do you think happens when equal concentrations of an acid and a base are mixed together? Why?

_____

_____

_____

_____

❸ If you mixed an acid into a base a little at a time, do you think you could tell whether the solution went from being more acidic to more basic? Why or why not?

_____

_____

_____

_____

CHEMISTRY

❹   What kinds of data do you think you could plot on a graph? Why?

_____

_____

_____

_____

_____

❺   What kinds of data do you think you could not plot on a graph? Why?

_____

_____

_____

_____

_____

❻   Do you think making a graph of data could be helpful? Why or why not?

_____

_____

_____

_____

_____

II. Experiment 4: Vinegar and Ammonia in the Balance: An Introduction to Titration   Date_____

Objective _____

_____

Hypothesis _____

_____

_____

Materials

    red cabbage indicator (from Experiment 3)
    household ammonia
    vinegar
    large glass jar
    measuring spoons
    measuring cup

EXPERIMENT

❶ Measure 60 ml (1/4 cup) of vinegar and put it in the glass jar.

❷ Add enough of the red cabbage indicator to get a deep red color.

❸ Carefully add 5 ml (1 tsp.) of ammonia to the vinegar solution. Swirl gently, and record the color of the solution in the chart on the following page.

❹ Add another 5 ml (1 tsp.) of ammonia to the vinegar solution, swirl the solution, and record the color.

❺ Keep adding ammonia to the vinegar solution 5 ml (1 tsp.) at a time. Record the color of the solution each time along with the total amount of ammonia that has been added.

❻ When the color has changed from red to green, add a few more ml (tsp.) of ammonia to see what happens.

CHEMISTRY

Results

Number of ml (tsp) of ammonia	Color of solution
5 ml (1 tsp)	red
10 ml (2 tsp)	red

Number of ml (tsp.) of Ammonia [continued]	Color of Solution [continued]

CHEMISTRY

❶ Plot the data from your chart using the graph on the next page. The horizontal axis is labeled **Amount of Ammonia Added**, and the vertical axis is labeled **Color of Solution**.

❷ For every 5 ml (1 tsp.) of ammonia added, mark the graph with a round dot corresponding to the color of the solution.

❸ When all of the data have been plotted, connect the dots.

Graphing Your Data

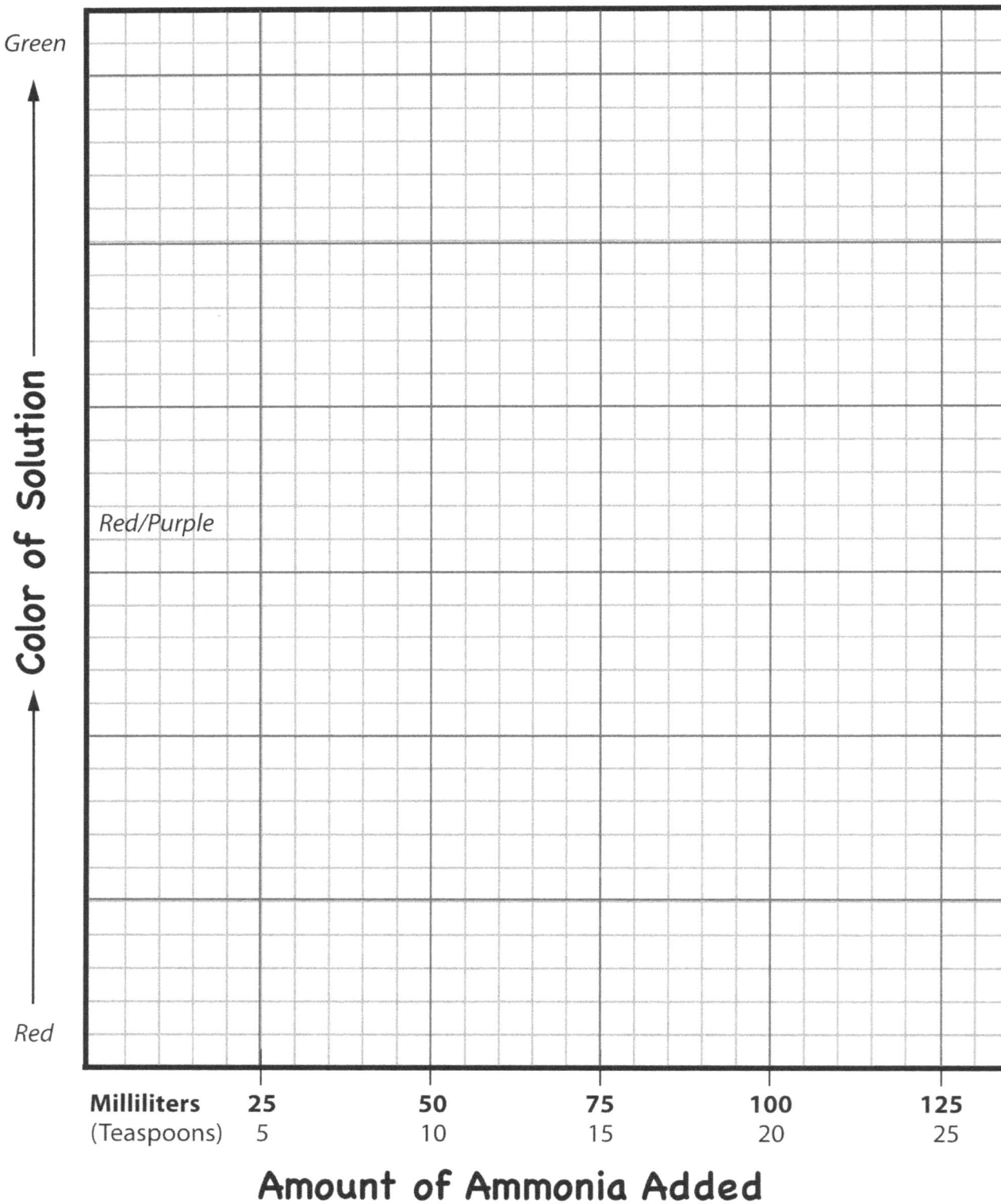

Green

CHEMISTRY

Color of Solution

Red/Purple

Red

| **Milliliters** | **25** | **50** | **75** | **100** | **125** |
| (Teaspoons) | 5 | 10 | 15 | 20 | 25 |

Amount of Ammonia Added

III. Conclusions

What conclusions can you draw from your observations?

_____

_____

_____

_____

_____

_____

_____

_____

_____

_____

_____

_____

_____

_____

_____

_____

_____

CHEMISTRY

IV. Why?

A titration is a technique in which one substance is added to another substance in small quantities. By adding the second substance in small quantities, any subtle change in the properties of the solution can be observed. The technique of titration is used not only for acid-base reactions but also for other types of reactions.

The concentration of something is simply the number of molecules in a given volume. A bottle of concentrated hand soap, for example, has more soap molecules (or less water) in it than one that is not concentrated. Because it is more concentrated, it takes less soap to make a lather than a less concentrated product, but it is still the same soap.

The same is true for acids and bases. If a solution is concentrated, it has more of the molecules that make it either acidic or basic. Therefore, it is stronger. This is illustrated with the difference between glacial acetic acid (concentrated acetic acid) and vinegar (dilute acetic acid). Glacial acetic acid has a very pungent odor and will burn skin on contact. Vinegar is the same acid, but much less concentrated and is safe to eat.

A neutral solution is one that is neither acidic nor basic. Acids and bases neutralize each other to make a salt and water. *Salt* refers to a product of an acid-base reaction rather than table salt.

For an acid-base reaction, the unknown concentration of an acid or base can be determined by doing a titration. How does this work? If you start with an acid solution and add a base, the base will neutralize the acid. Once all of the acid has been neutralized, the next drop of base will cause the pH to change dramatically. If the concentration of one solution (acid or base) is known, then the concentration of the other, unknown solution can be determined.

The neutralization process of acids and bases is used in everyday life. For example, antacids are bases that are used to neutralize hydrochloric acid in the stomach when someone has acid indigestion.

V. Just For Fun

Find some household solutions that you think may be an acid or a base. Test them with the red cabbage juice indicator. Choose one acid and one base and do a titration, following the steps of the previous experiment. Charts are provided for recording your data and plotting your graph.

Acid _____

Base _____

Number of ml (tsp.) of base added	Color of solution

CHEMISTRY

Number of ml (tsp.) of base added	Color of solution

Graphing Your Data

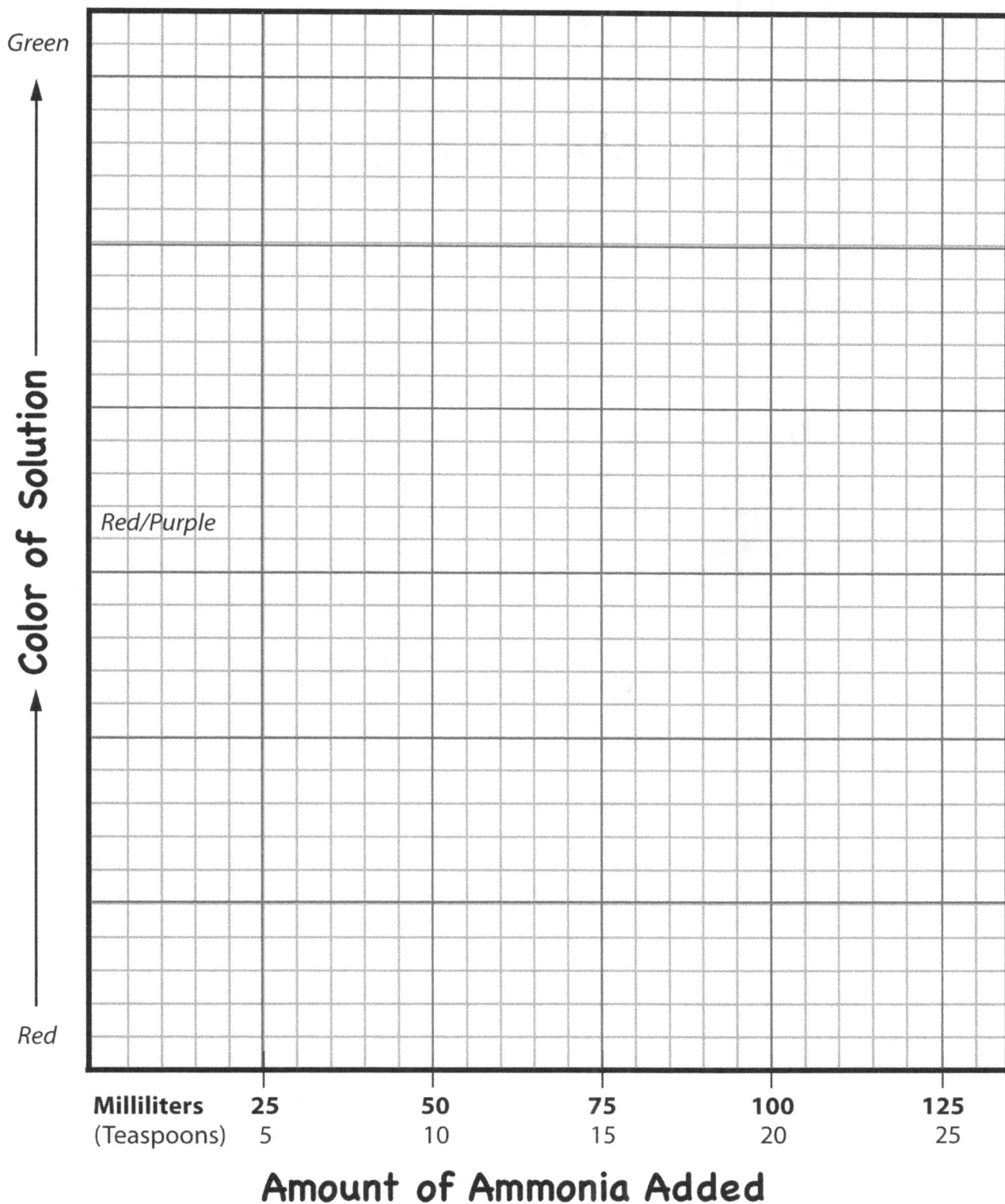

Green

Color of Solution

Red/Purple

Red

| Milliliters | 25 | 50 | 75 | 100 | 125 |
| (Teaspoons) | 5 | 10 | 15 | 20 | 25 |

Amount of Ammonia Added

CHEMISTRY

Experiment 5

Show Me the Starch!

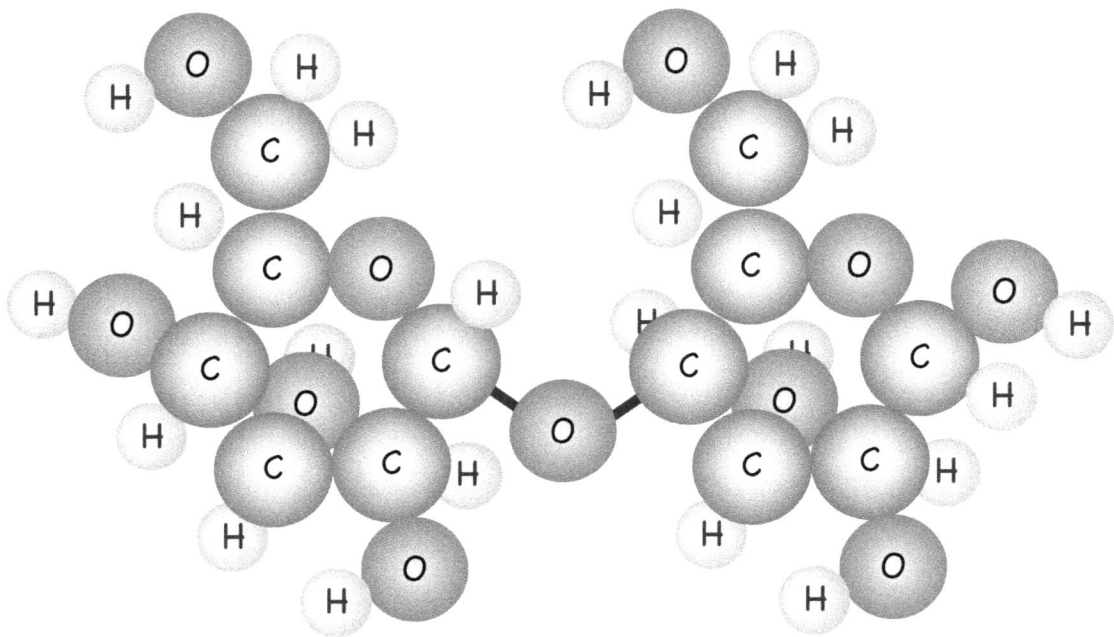

Introduction

Explore the presence and absence of starches in food.

I. Think About It

❶ Do you think there are different kinds of sugar? Why or why not?

_____

_____

_____

_____

_____

❷ Do you think all starches in food are the same? Why or why not?

_____

_____

_____

_____

_____

❸ Do you think your body can use both cellulose and starches for food energy? Why or why not?

_____

_____

_____

_____

_____

CHEMISTRY

❹ Why do you think horses and people eat different kinds of plants?

_____

_____

_____

_____

_____

❺ Do you think all carbohydrates are the same? Why or why not?

_____

_____

_____

_____

_____

❻ Do you think bacteria are important for the digestion of food? Why or why not?

_____

_____

_____

_____

_____

II. Experiment 5: Show Me the Starch!   Date _____

Objective _____

_____

Hypothesis _____

_____

_____

_____

Materials

tincture of iodine **[Iodine is VERY poisonous—DO NOT EAT** any food items
with iodine on them.]
a variety of raw foods, including:
pasta
bread
celery
potato
banana and other fruits
liquid laundry starch (or mix equal parts borax and corn starch into water)
absorbent white paper
eye dropper
cookie sheet
marking pen

EXPERIMENT

❶ Take several food items and place them on a cookie sheet.

❷ Using the eye dropper, put a small amount of liquid starch on a piece of
absorbent paper and label it **Control**. Let it dry.

❸ Add a drop of iodine to the starch on the control paper. Record the color in
the following chart.

❹ Add iodine to each of the food items and record the color for each.

❺ Compare the color of the **Control** to the color of each food item.

❻ Note those food items that changed color and those that did not.

Results

Food Item	Color
Control:	

III. Conclusions

What conclusions can you draw from your observations?

_____

_____

_____

_____

_____

_____

_____

_____

_____

_____

_____

_____

_____

_____

_____

_____

_____

_____

CHEMISTRY

IV. Why?

Carbohydrates are energy producing molecules. The largest carbohydrates are the polysaccharides, which include the starches and cellulose. Amylose and amylopectin are the two main starches found in potatoes, breads, and pasta, and these two main energy molecules found in plants can be used by human bodies for food.

In this experiment it is amylose that is detected by iodine. Amylose interacts with the iodine molecules to give a deep purple or black color. Amylopectin cannot be detected by iodine because it doesn't interact with the iodine molecules.

Cellulose is also a polysaccharide and is the main structural molecule for plants. Because plants don't have bones, they need cellulose to provide a rigid structure for support.

The structure of cellulose is very different from that of starches. The main difference between the bonding in starches and in cellulose is the orientation of the bond between individual glucose units. For the starches, the bond points down and for cellulose the bond points up. This small difference in how the units are hooked to each other makes a huge difference in their overall shape. The long chains in starches are either coils or large branched molecules. In contrast, cellulose molecules form large sheets which can stack on one another. The cell walls in plants are made of layers of parallel cellulose sheets that make the cell wall rigid.

Although humans can get many nutrients from some plants, our bodies cannot get energy molecules from the part made of cellulose. We do not have the enzymes required to break the bonds of the glucose units that form cellulose. Certain animals have bacteria that produce enzymes to break the cellulose bonds, enabling them to get glucose from a diet that consists of grasses and other plants.

V. Just For Fun

Find out how a banana changes over time.

Slice a piece off a green banana. Test it with a drop of iodine.

Every third day slice another piece off the banana and test it with iodine.

Record the results of each test.

Date	Observations

Experiment 6

Using Agar Plates

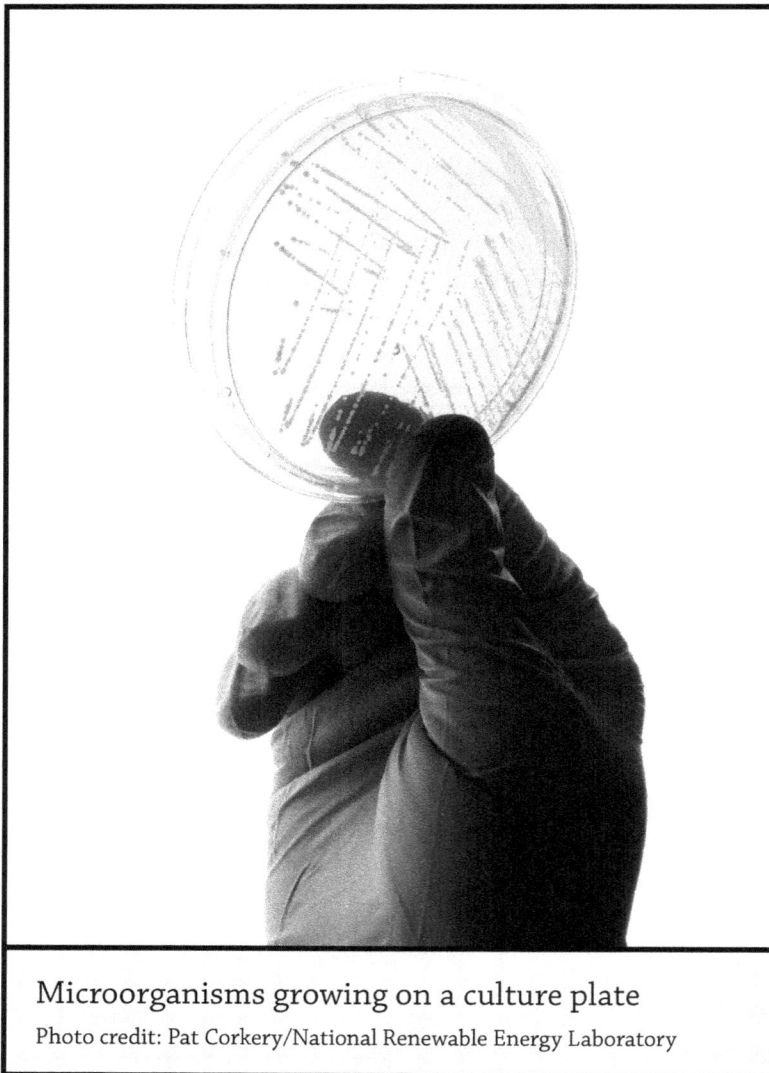

Microorganisms growing on a culture plate

Photo credit: Pat Corkery/National Renewable Energy Laboratory

Introduction

In this experiment you will be culturing bacteria—growing bacteria in agar, a specially prepared substance made from seaweed that contains nutrients needed for growth. For the bacteria to grow, it also needs to be incubated, or kept under conditions that are favorable for growth. The petri dishes prepared with agar are called plates.

I. Think About It

❶ What do you think will happen to an agar plate if it is not cooled correctly?

_____

_____

_____

_____

_____

_____

❷ What do you think will happen if there are bubbles in the agar?

_____

_____

_____

_____

_____

_____

BIOLOGY

❸ What do you think will happen if the agar does not gel correctly?

_____

_____

_____

_____

_____

_____

❹ What do you think will happen if an agar plate is old and starts to dry up?

_____

_____

_____

_____

_____

_____

❺ What do you think will happen if an agar plate gets wet after a bacteria sample has been added?

_____

_____

_____

_____

_____

_____

II. Experiment 6: Using Agar Plates

Date _____

Objective _____

_____

Hypothesis _____

_____

_____

BIOLOGY

Materials

- plastic petri dishes
- dehydrated agar powder
- distilled water
- safe *E. coli* bacterial culture
- inoculation loop
- candle or gas flame
- cooking pot
- mixing spoon
- oven mitt or pot holder
- measuring spoons
- measuring cup
- black permanent marker
- red marker
- rubber gloves

EXPERIMENT

Part I: Preparing Agar Plates

❶ Being careful not to remove the lids, spread the petri dishes out on a clean countertop.

❷ Put 200 ml (about 1 cup) room temperature distilled water in a cooking pot and add 10 ml (2 teaspoons) dehydrated agar powder. Bring to a boil while stirring.

❸ Using an oven mitt or pot holder, hold in one hand the cooking pot containing the hot agar. With the other hand gently slide the petri dish lid to the side and pour in enough hot agar to just cover the bottom. Carefully slide the petri dish lid back on to cover the agar. Repeat for all the petri dishes.

As you pour each plate, observe whether the agar surface has bubbles. If it does, use a marker to put a red dot on the lid. If you are nearly done with pouring the plates and have no plates with bubbles, gently shake the agar while pouring it to create bubbles.

❹ Allow the agar in the petri dishes to cool completely. It will form a hard surface.

❺ Once cooled, separate those plates that have bubbles in the agar from those plates that do not have bubbles. Mark the lids of the plates as follows:

Take one of the plates that has bubbles and turn it upside down, agar side up. Mark it "Bubbles" and store it agar side up in the refrigerator.

Take a plate without bubbles and turn it upside down, agar side up. Mark it "Up" and store it agar side up in the refrigerator.

Take a plate without bubbles and leave it agar side down. Mark it "Down" and store it agar side down in the refrigerator.

Take a plate with or without bubbles, remove the lid, and allow it to dry until the edges of the agar start to pull away from the sides of the petri dish. This may take a few days depending on the humidity in your region. Replace the lid and mark it "Dry." Store right side up at room temperature.

❻ Store the remaining plates upside down in the refrigerator until you are ready to use them.

Part II: Streaking the *E. coli* Culture

Wear rubber gloves during this experiment.

❶ Take the inoculating loop and sterilize it by heating it in a gas flame or in a candle flame until the wire turns orange. [Note: a candle flame is hot enough to sterilize the loop, but keep the loop close to the blue part of flame where it is cleaner and has less soot.]

❷ When the loop turns orange, remove it from the flame and allow it to cool completely. To avoid contamination, the sterilized loop should not touch anything while it is cooling.

❸ Dip the loop in the inoculating *E. coli* culture and remove the loop from the tube without touching the sides of the tube. To avoid contamination, the loop should not touch the sides of the tube while you are inserting and removing it.

Repeat this step for each plate.

❹ Streak all the marked plates by using a zigzag motion from one side of the plate to the other. Store the plates agar side up at room temperature.

❺ Use one of the plates that does not have bubbles and was stored in the refrigerator agar side up. Pour enough *E. coli* culture to cover the agar and replace the lid on the petri dish. Mark this plate "Spread." Allow the agar to absorb the *E. coli* and then store the plate agar side up at room temperature with the other plates.

Results

After your *E. coli* cultures have incubated for 1–3 days, carefully inspect each plate. Notice whether any dots (*E. coli* colonies) are visible on the plates and if so, note how big they are. In the following chart, record your observations including comparisons of the different plates to each other.

BIOLOGY

Plate Description	Observations
Bubbles	
Up	
Down	
Dry	
Spread	

III. Conclusions

How well do you think the plates performed while growing bacteria? Do you think some plates worked better than others? Why or why not?

_____

_____

_____

_____

_____

_____

_____

_____

_____

_____

_____

_____

_____

_____

_____

_____

_____

BIOLOGY

IV. Why?

Preparing and using agar plates for growing colonies of bacteria or other organisms can be an art. Good agar plate preparation will produce plates with no bubbles, the right amount of solid agar reaching the edges of the petri dish, and no condensed water being present. Also, to do bacterial experiments successfully, all of the plates must be the same.

In this experiment you were able to compare what happens to plates that have bubbles, whether or not the way they are stored caused water condensation on the agar, and what happens when a plate is dried out. All of these issues may cause problems for growing good quality bacterial colonies. Bubbles may cause problems with identifying colonies depending on how big the bubbles are. If a bubble pops, a depression in the agar can form, pooling the bacterial culture which might cause it to grow in an uncontrolled way. If the agar breaks during the streak, bacteria may grow inside the agar in an uncontrolled way. Water condensation on an agar plate will often cause irregular bacterial growth because water can dilute the nutrients in the agar, causing the nutrients to vary from plate to plate. Dry plates often don't work for growing good colonies because the concentration of nutrients may change, and there may not be enough water moisture for controlled growth. Plates with any of these issues are typically discarded.

Also, you observed what happens when a bacterial culture is streaked vs. spread on an agar plate. Streaking a bacterial culture on a plate will usually result in individual colonies that look like separate dots. Being able to isolate single colonies is important for microbiological and genetic research. Since more of the bacterial culture is left on the agar at the beginning of the streak than the end, you may notice a larger area of bacteria that is a clump of colonies at the beginning of the streak and single colonies (dots) at the end of the streak. If the researcher suspects that the bacterial count is low, meaning there are very few bacteria in solution, spreading a given volume on a plate rather than streaking it will help make sure that some bacteria will be on the plate.

BIOLOGY

V. Just For Fun

Take several agar plates and see what happens if you use a different streaking pattern on each one. What happens if you make circles on the plate? What happens if you use a crosshatch pattern? What happens if you streak different letters on the plate like a T, O, P or X? After allowing the cultures to incubate until dots are visible (about 1-3 days), examine the plates and record your results.

Streak Pattern	Observations

BIOLOGY

Experiment 7

Using a Light Microscope

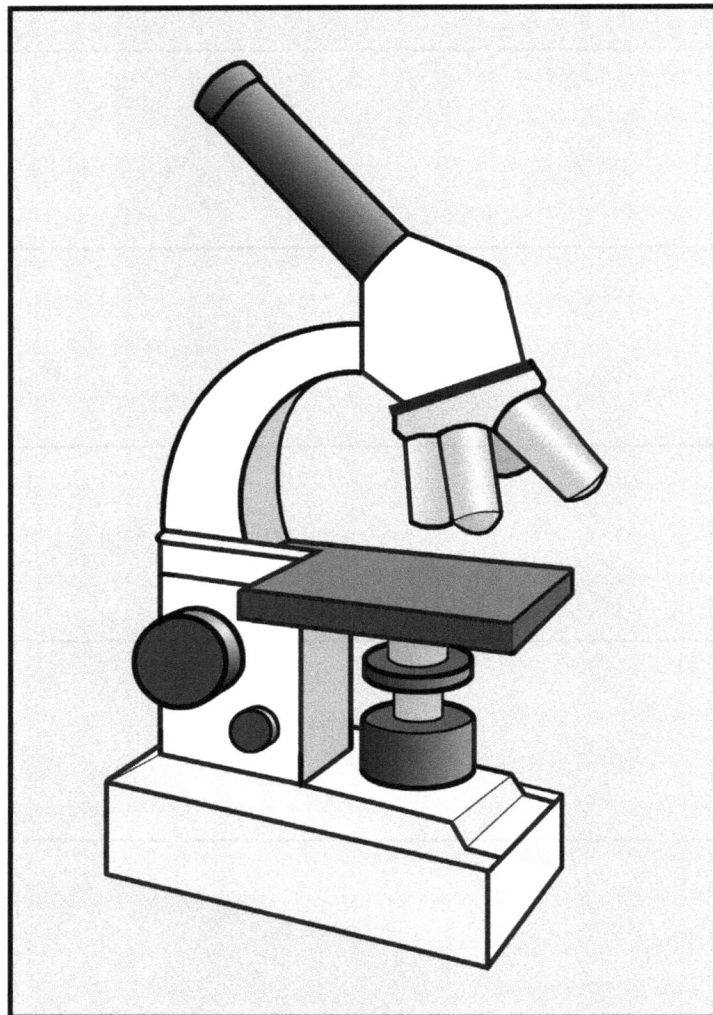

Introduction

Explore the world through a microscope!

I. Think About It

❶   What more do you think you could observe about a strand of your hair by looking at it through a microscope instead of with just your eyes?

_____

_____

_____

_____

❷   What do you think a strand of hair would look like at 40X magnification?

_____

_____

_____

_____

❸   What do you think a piece of paper would look like at 10X magnification?

_____

_____

_____

_____

❹   How easy do you think it is to get a sample in focus at 10X magnification?

_____

_____

_____

_____

_____

❺   What do you think pond water would look like at 10X magnification?

_____

_____

_____

_____

_____

❻   What do you think skin cells would look like at 100X magnification?

_____

_____

_____

_____

_____

BIOLOGY

II. Experiment 7: Using a Light Microscope   Date _____

Objective _____

_____

Hypothesis _____

_____

_____

Materials

Microscope with 4X, 10X, and 40X objective lenses. A 100X objective lens is recommended but not required.

glass microscope slides

glass microscope cover slip

immersion oil (if using 100X objective lens)

Samples:

- piece of paper with lettering
- strands of hair
- droplet of blood
- insect wing

EXPERIMENT

Part I: The Microscope

❶ Move the microscope to a desk or table where you can sit for a few hours.

❷ Remove the cover and familiarize yourself with the different parts.

❸ Turn the revolving turret so the lowest power objective lens (4X) is clicked into position. **[NOTE: Be extremely careful not to bang the lenses on the stage as you turn the turret. This can damage the lenses.]**

❹ Turn the coarse adjust knob and observe how the stage moves up and down.

❺   Turn the fine adjust knob and observe how the stage moves up and down.

❻   Turn on the light source and use the condenser to change the amount of light entering the stage.

❼   Examine the stage and the clips that hold the slide in place. Move them until you are familiar with how they work.

❽   Fill in the names of the parts of the microscope in the following diagram.

BIOLOGY

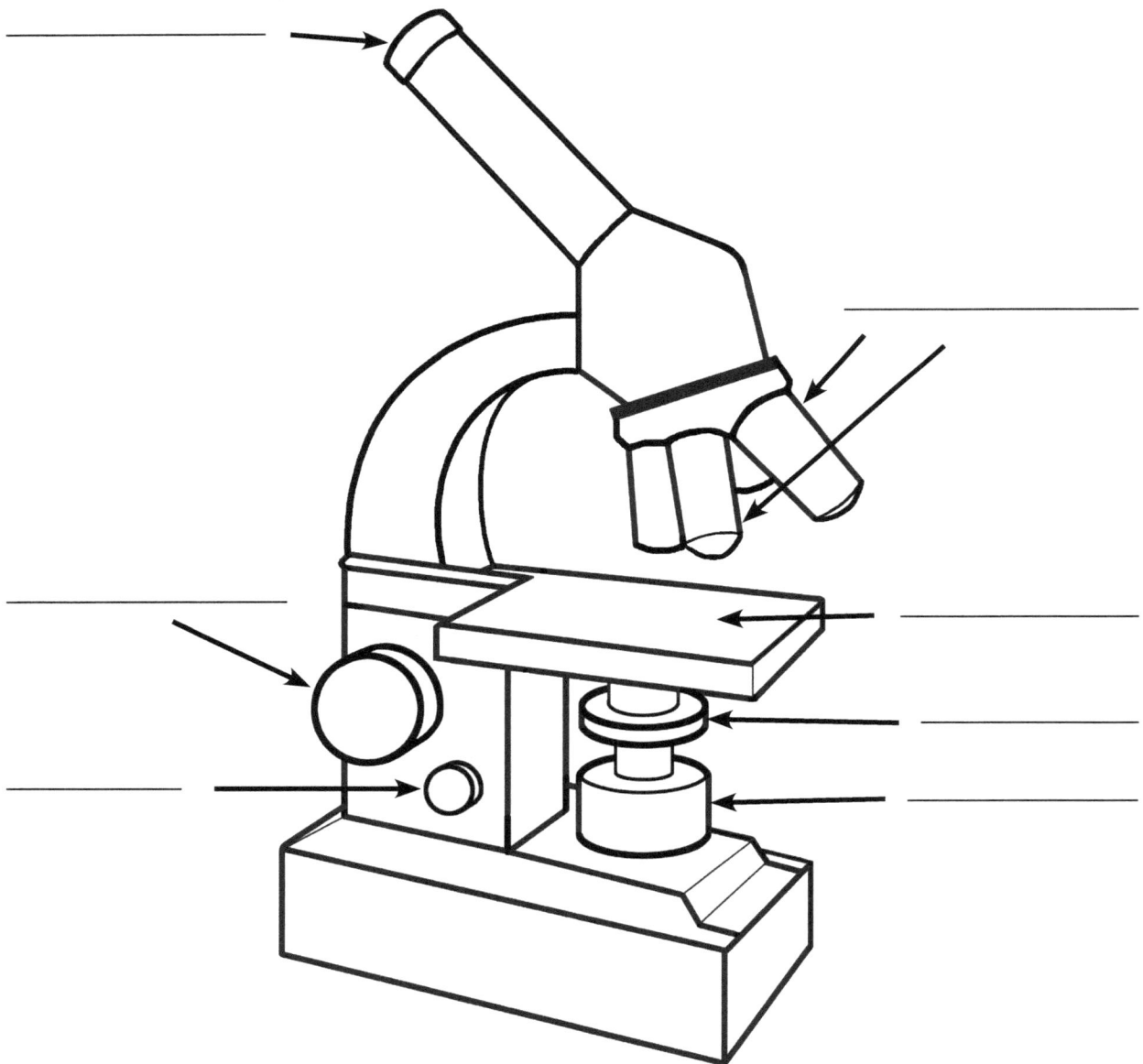

Part II: Observing a Sample

1. Take the small piece of paper and place it on a glass slide in the microscope. Do not put a coverslip on top.

2. With the lowest power objective (4X) in place, look through the ocular lens, and using the coarse adjust knob, slowly move the objective lens up and down until the sample is in focus.

3. Move the paper until you can see the ink.

4. Turn the fine adjust knob slowly up and down as you observe the sample. You should see some parts of the paper come into focus as other parts go out of focus. Notice the range of focus (how much of the sample is in focus). Record your observations.

_____

_____

_____

_____

5. Keeping the microscope steady, gently turn the turret until the next highest power objective lens (10X) clicks into place. Be careful not to bump the objective lens into the sample. If this looks like it will occur, move the lens up with the coarse adjust knob.

6. Look through the ocular lens and observe the sample, turning the coarse and then fine adjust knobs until the sample comes into focus. Notice the range of focus. Record your observations.

_____

_____

_____

_____

BIOLOGY

7. Using the turret, move the lowest power objective back into place. [**NOTE:** If you have a 100X objective, do not rotate the turret through this lens. Turn the turret in the opposite direction until the lowest power lens is back in place. **It is extremely important that the lens does NOT scrape the slide or sample. This will scratch the lens and ruin it.**]

8. Take a clean glass slide and place a single droplet of blood on the slide. You can either let a drop of blood fall on the slide or touch your finger to the slide to transfer the blood. You will need to prick your finger or the finger of a brave sibling, parent, or friend. Hands should be washed first and a sterile needle used.

9. Gently place a clean coverslip on the droplet. The blood droplet should spread out quickly when the coverslip is in place.

10. Put the glass slide in the sample holder, and with the 4X objective in place, look through the ocular lens. Focus the image with the coarse and then fine adjust knobs. Record your observations and draw what you see.

BIOLOGY

11. Gently turn the turret until the 10X objective clicks in place. Rotate the lens into place slowly, being careful not to bump the slide with the objective lens. Record your observations and draw what you see.

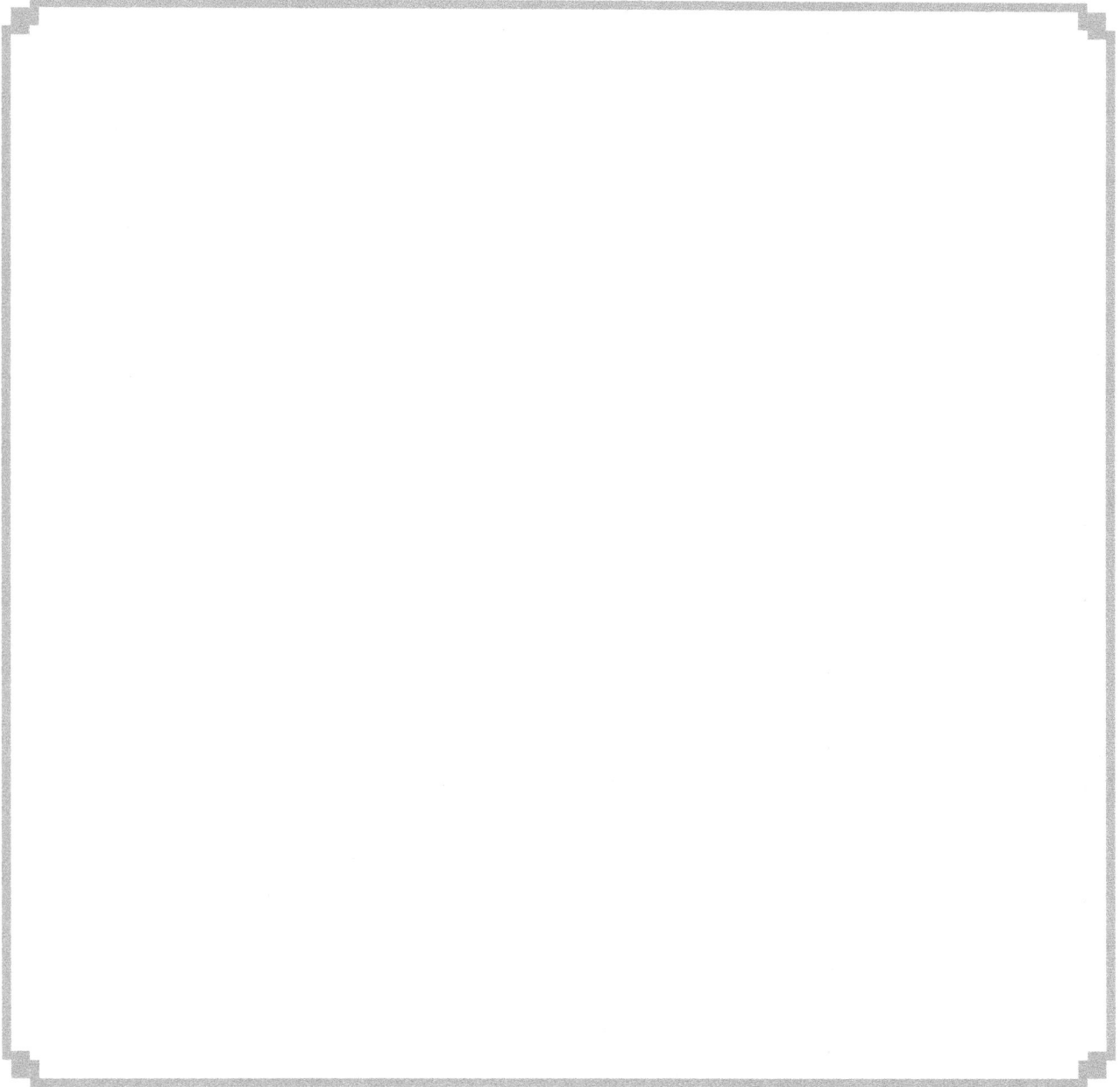

12. Gently turn the turret until the 40X objective clicks into place. Again, rotate the lens into place slowly and do not bump the slide with the objective lens. There should be enough room to move the lens without moving the stage. Record your observations and draw what you see.

BIOLOGY

13. If you have a 100X oil immersion lens, turn the turret until the oil immersion lens is half way into position. Place a single drop of immersion oil on the glass cover slip and gently move the oil immersion lens in place.

[NOTE: It is extremely important that the lens does NOT scrape the coverslip. This can scratch the lens and ruin it. The lens will be very close to the glass coverslip if you have adjusted the focus correctly with the other lenses. If it seems like the lens will scrape the cover slip, gently back up the lens with a few turns of the fine adjust knob.]

14. Adjust the condenser if more light is needed, and using the fine adjust knob only, focus the image. **[Never use the coarse adjust knob to focus an oil immersion lens—it is too easy to smash the lens against the slide.]** Record your observations and draw what you see.

15. Rotate the turret to move the oil immersion lens away from the sample. With the coarse adjust knob lower the stage and remove the sample.

16. Explore other samples such as a strand of hair, pond water, or an insect wing. Always rotate the lenses from lowest power to highest power, focusing the image each time before moving to the next highest power lens.

III. Conclusions

What conclusions can you draw from your observations?

_____

_____

_____

_____

_____

_____

_____

_____

_____

_____

_____

_____

_____

_____

_____

_____

BIOLOGY

IV. Why?

In this experiment you explored how to operate a compound light microscope. Being able to select and then focus the correct objective lens for viewing different types of samples is an important first step towards becoming a good microscopist.

In the first part of the experiment, you magnified a piece of paper with a 4X objective, which is usually the lowest power objective lens on a standard student microscope. When combined with a 10X ocular lens, a 4X objective magnifies the sample 40X. At this magnification you can begin to see structures in paper that you can't see with your eyes alone. Moving the 10X objective lens into place magnifies the paper 100X, allowing more details to be observed.

As you increase the power of the objective lens, the depth of field decreases. In other words, you may have noticed that the focal length decreases as the magnification increases. When an objective lens is made with good optics, the working distance, or focal length, decreases as the power of the objective increases and as the resolution increases.

You may have also noticed that the field of view (the area of the sample that is visible) decreases as the magnification increases. As you move from a low power objective lens (4X) to a higher power objective lens (40X), the field of view decreases.

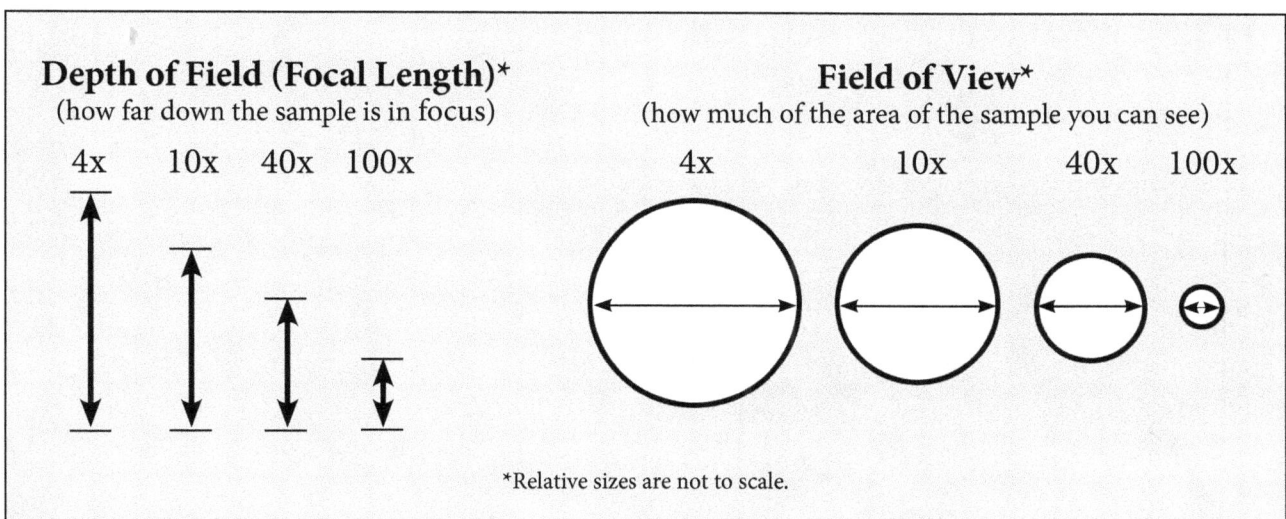

Depth of Field (Focal Length)*
(how far down the sample is in focus)

4x   10x   40x   100x

Field of View*
(how much of the area of the sample you can see)

4x   10x   40x   100x

*Relative sizes are not to scale.

BIOLOGY

The change in the field of view and in the focal length as you move from lower to higher magnification can be one of the most frustrating experiences for new microscopists. For this reason, it is important to focus and center the sample beginning with the lowest power lens, then working up to the highest magnification, focusing and centering the sample with each change of lens. If the sample is bumped or moves out of view while viewing with a high magnification lens, it is good practice to go back to a lower power objective and refocus and re-center the sample. However, if an oil immersion lens is used, often the area being viewed needs to be adjusted or a new sample created and the process of going from lowest to highest magnification repeated.

V. Just For Fun

Now that you know how to use a microscope, the world of tiny objects and microscopic organisms is at your fingertips.

Using your microscope, observe what milk, soda, clothing fibers, skin, fingernails, or other samples look like when magnified. Does your hair look the same as your dog's or cat's or horse's? Do fibers from different clothing materials look the same? Which samples were the most surprising when magnified? Record your observations. Draw several of your favorite samples as seen through the microscope.

Observations Through a Microscope

More Observations Through a Microscope

Experiment 8

Observing Protists

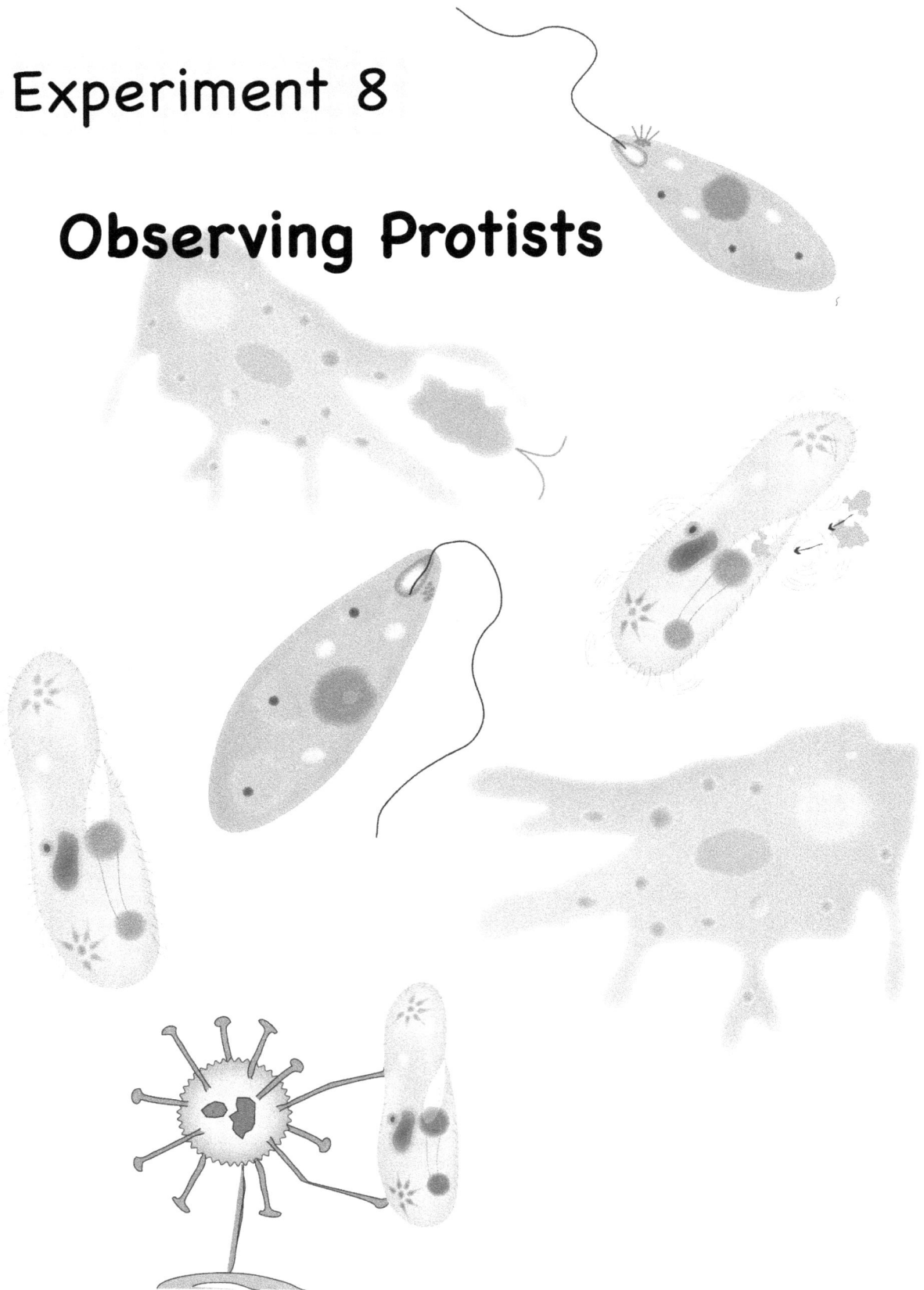

Introduction

Look through a microscope to see some amazing protists!

I. Think About It

❶ How do you think protists move? Do they all move in the same way? Why or why not?

_____

_____

_____

_____

❷ Would you rather be able to move like an amoeba or a paramecium? Why?

_____

_____

_____

_____

❸ What do you think it would be like to have to eat like an amoeba does?

_____

_____

_____

_____

❹ What would be the differences between moving with cilia and moving with a flagellum? Why?

_____

_____

_____

_____

BIOLOGY

❺ Do you think it would be easier for a euglena to eat than for a paramecium to eat? Why or why not

_____

_____

_____

_____

❻ In what ways do you think protists are like plants and how are they like animals?

_____

_____

_____

_____

II. Experiment 8: Observing Protists—How Do They Move?

Date _____

Objective _____

_____

Hypothesis _____

_____

_____

Materials

microscope with a 10X objective
microscope depression slides
5 eyedroppers
fresh pond water or water mixed with soil
protist study set
methyl cellulose

EXPERIMENT—Part A

❶ Take one of the samples from the protist set and use an eyedropper to place a small droplet of solution onto a glass slide that has been correctly positioned in the microscope.

❷ Observe the movement of the protists. If the organisms move too quickly, apply a droplet of methyl cellulose into the protist solution on the slide.

❸ Patiently observe the movement of one type of protist. In the *Results— Part A* section, note the type of protist you are observing. Draw the protist and describe how it moves, writing down as many observations as you can.

❹ Repeat Steps ❶-❸ with two other protist types. Use a new eyedropper for each sample.

Results—Part A

Type of protist _____

Observations of movement:

_____

_____

_____

_____

_____

_____

Drawing of the protist

Type of protist _____

Observations of movement:

_____

_____

_____

_____

_____

_____

_____

Drawing of the protist

BIOLOGY

Type of protist _____

Observations of movement:

_____

_____

_____

_____

_____

_____

Drawing of the protist

EXPERIMENT—Part B

Take a droplet of fresh pond water (or water mixed with soil) and place it on a slide in the microscope. Based on how the organisms move, try to determine the types of protists you are observing. Write and draw your observations in the *Results—Part B* section.

Results—Part B

Types of protists seen and observations about them

_____

_____

_____

_____

_____

_____

_____

_____

_____

_____

_____

_____

_____

_____

BIOLOGY

Drawings of the protists labeled with type of protist

BIOLOGY

More protist observations

III. Conclusions

What conclusions can you draw from your observations?

_____

_____

_____

_____

_____

_____

_____

_____

_____

_____

_____

_____

_____

_____

_____

_____

BIOLOGY

IV. Why?

Protists are members of the kingdom Protista, and they are found wherever there is water, including saltwater, freshwater, and soil. Protists have been difficult to classify because they are eukaryotes and can have both plant-like and animal-like qualities. Protist is an "umbrella term" that fits those organisms that cannot be easily placed in any other kingdom.

The euglena moves by whipping back and forth an appendage called a flagellum, and the paramecium uses hair-like projections called cilia that it beats to move itself around. Although they have a simple appearance in a microscope, cilia and flagella are actually very sophisticated machines. Each whip contains long strands of molecules called microtubules. As the microtubules slide past each other, the flagellum or cilium changes orientation. When the microtubules next slide past each other going the opposite way, the whip again changes orientation. These successive changes cause the cilia or flagella to beat or whip causing the protist to move.

Having one flagellum makes a euglena tend to move in a single direction, and it may hover under a light source as it gathers light for photosynthesis. Having many cilia allows a paramecium to move all over the place. It can roll, move forward and backward, and spin. Amoebas move the most slowly, as they expand and contract their pseudopods to crawl along.

Protists are uniquely designed and are amazing tiny single-celled creatures.

BIOLOGY

V. Just for Fun: How Do They Eat?

Date: _____

Do another experiment, this time observing how protists eat.

Objective _____

_____

Hypothesis _____

_____

Materials

microscope with a 10X objective
microscope depression slides
5 eyedroppers, measuring cup and measuring spoons
protist study kit
baker's yeast
distilled water
Eosin Y stain

EXPERIMENT

❶ Color the yeast with Eosin Y stain as follows:

Add 5 milliliters (one teaspoon) of dried yeast to 120 milliliters (1/2 cup) of distilled water. Allow it to dissolve.

Add one droplet of Eosin Y stain to one droplet of yeast mixture on a slide. Look at the mixture under the microscope. You should be able to see individual yeast cells that are stained red.

❷ Get the amoeba sample and place a small droplet of the solution onto a glass slide that has been correctly positioned in the microscope.

❸ Take a small droplet of the yeast stained with Eosin Y and place it into the droplet of protist solution that is on the slide.

❹ Looking through the microscope, patiently observe the protists, and note the red-colored yeast. Look for a protist eating and try to describe how it eats. In the *Results* section write down as many observations as you can. Draw one of the protists eating.

❹ Repeat steps 2-4 using the paramecium sample.

Results

Observations of an amoeba eating

_____

_____

_____

_____

_____

_____

_____

Drawing of how an amoeba eats

BIOLOGY

Observations of a paramecium eating

_____

_____

_____

_____

_____

_____

_____

Drawing of how a paramecium eats

Conclusions

What conclusions can you draw from your observations?

_____

_____

_____

_____

_____

_____

_____

_____

_____

_____

_____

_____

_____

_____

_____

_____

Experiment 9

Moldy Growth

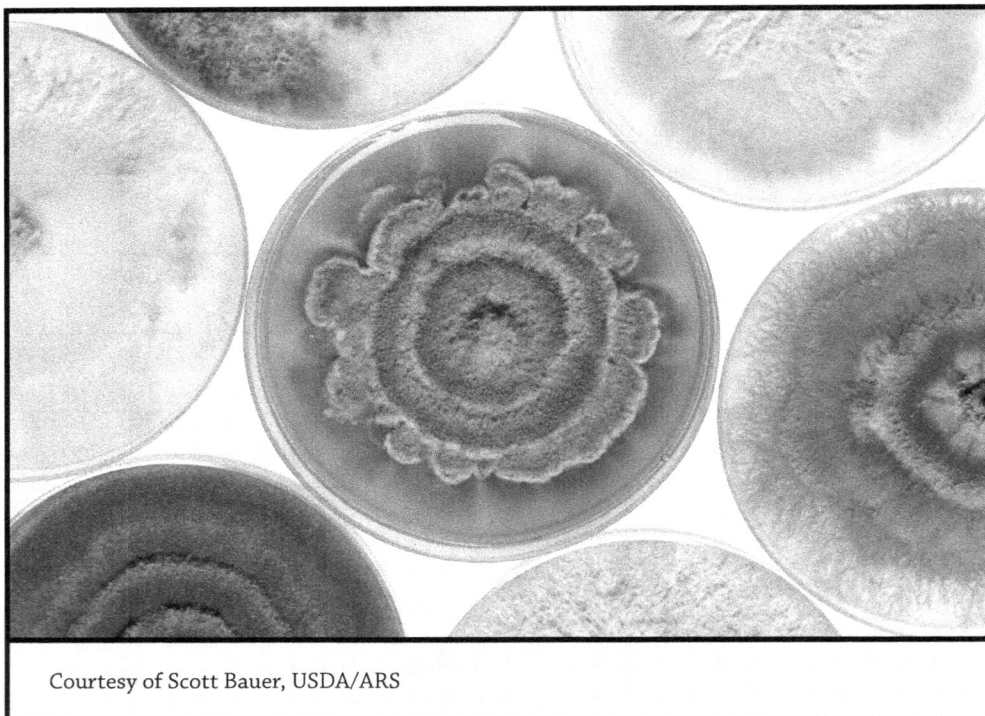

Introduction

Find out which household products will control the growth of mold.

I. Think About It

❶ Have you seen mold growing in your shower? What did it look like?

_____

_____

_____

_____

_____

❷ Have you ever seen mold on bread? What did it look like?

_____

_____

_____

_____

_____

❸ How do you think you could get rid of mold that is growing in the shower?

_____

_____

_____

_____

❹ What cleaning products do you think might kill mold?

_____

_____

_____

_____

_____

❺ Do you think a different kind of fungus such as a mushroom could grow in your shower? Why or why not?

_____

_____

_____

_____

❻ Do you think molds could make bread rise like yeast does? Why or why not?

_____

_____

_____

_____

_____

II. Experiment 9: Moldy Growth

Date _____

Objective _____

_____

Hypothesis _____

_____

_____

Materials

dehydrated agar powder
distilled water
cooking pot
measuring spoons
measuring cup
plastic petri dishes
permanent marker
oven mitt or pot holder
jar with lid (big enough to hold 235 ml (about 1 cup) liquid
1 slice of bread, preferably preservative free
small clear plastic bag
white vinegar
bleach
borax
mold or mildew cleaner
rubber gloves

EXPERIMENT

Part I: Pour Agar Plates

Follow the directions in **Experiment 6** for making agar plates. You will need six plates that do not have bubbles in the agar for this experiment, and at least one more for the *Just For Fun* experiment. Store them upside down (agar side up) in the refrigerator until you are ready to use them.

Part II: Observing Mold

❶ Take a slice of bread, place it in a clear plastic bag, and add 5 ml (1 teaspoon) of water. Seal the bag and place it in a dark, warm area for several days until mold is visible on the bread.

❷ Put on rubber gloves and cut 5 small cubes from the moldy bread. Put the rest of the bread back in the plastic bag to save for the *Just For Fun* experiment.

❸ Take one agar plate and mark it "Control."

❹ Take five more agar plates and mark them "None," "Vinegar," "Bleach," "Borax" and "Mildew Cleaner." Place the plates agar side down.

❺ To plate "Vinegar" add 5 ml (1 tsp.) of white vinegar and spread it evenly on the surface of the agar by gently tilting the plate back and forth.

❻ To plate "Bleach" add 5 ml (1 tsp.) of bleach and spread it evenly on the surface of the agar by gently tilting the plate back and forth.

❼ To plate "Borax" add 5 ml (1 tsp.) of borax to 235 ml (about 1 cup) of distilled water in a jar and dissolve completely by shaking. Then measure 5 ml of this mixture and spread it evenly on the surface of the plate by gently tilting the plate back and forth.

❽ To plate "Mold Cleaner" add 5 ml (1 tsp.) of mildew cleaner and spread it evenly on the surface of the agar by gently tilting the plate back and forth.

❾ To each plate except "Control" add a small cube of moldy bread and close the lid. The plate labeled "None" will only have the moldy bread and no cleaner added. The plate marked "Control" will contain only the agar and will have nothing added to it.

❿ Move all the plates to a dark room that is optimally 27°C (80°F). Store it agar side down

BIOLOGY

Results

Observe the plates for several days. Write and draw your observations in the following table.

Plate	Observations
Control	
None	
Vinegar	
Bleach	
Borax	
Mold Cleaner	

III. Conclusions

What conclusions can you draw from your observations?

_____

_____

_____

_____

_____

_____

_____

_____

_____

_____

_____

_____

_____

_____

_____

_____

BIOLOGY

IV. Why?

You may have noticed mold growing on your bathroom tiles or in the corners of your kitchen. This is because molds grow in warm, dark, and moist environments. Mold is everywhere, even though you can't always see it. Mold spores are microscopic in size and can be in the air or on surfaces and you will never see them! Mold spores can stay in a dormant, or resting, state until conditions are favorable for growth, at which time hyphae begin to grow. Molds live by digesting organic matter, and therefore they can damage wooden building materials and furniture. Because molds require moisture to grow, one way to control the growth of mold is to keep surfaces dry.

Mold growth can be also be controlled by using different cleaners and natural substances. Most molds like to be in a somewhat acidic environment. Some natural substances, such as borax, are high in pH and as a result create a basic environment for mold that is unfriendly. Borax has an extremely basic pH of 9.3 which is high enough kill both mold mycelia and spores. Borax is able to get below the surface of porous materials where it will kill mold mycelia. It can be an effective and safe bathroom cleaner and can be used to prevent mold from growing on freshly cut wood.

Vinegar has an acidic pH and kills about 82% of molds. It is not generally as effective at killing and preventing the growth of mold as Borax is. Bleach has a basic pH and is useful for killing bacteria but generally kills only molds that are growing on nonporous surfaces. Bleach is not effective for reaching the mycelia of molds growing on porous surfaces.

V. Just For Fun

Part I

Repeat the experiment using 3% hydrogen peroxide. How do these results compare? If you have some extra prepared plates, find some other household products to test. Record your results.

Part II

Look at the moldy bread through your microscope. Record your results.

Part I. Hydrogen Peroxide Results

Part II. Bread Mold Up Close

BIOLOGY

Experiment 10

Using Electronics

Introduction

Find out how electric circuits work!

I. Think About It

❶ How many modern toys, tools, and appliances can you think of that have electric circuits?

_____

_____

_____

_____

❷ What do you think an electric circuit looks like? Draw your idea here.

PHYSICS

❸ What do you think the difference is between an electric circuit and an integrated circuit?

_____

_____

_____

❹ Do you think integrated circuits include electric circuits? Why or why not?

_____

_____

_____

_____

_____

❺ What do you think an integrated circuit looks like? Draw your idea here.

PHYSICS

❻ How do you think toys, tools, and appliances with integrated circuits are different from those that have only electric circuits? Why?

_____

_____

_____

_____

II. Experiment 10: Using Electronics   Date _____

Objective _____

_____

Hypothesis _____

_____

_____

Materials

An electronic circuit kit

Recommended kits:

Snap Circuits: http://www.snapcircuits.net/
Snap Circuits Jr. 100 Experiments Kit

Little Bits: http://littlebits.cc/intro
Base Kit: http://littlebits.cc/kits/base-kit

PHYSICS

EXPERIMENT

❶ Open the electronic kit and study the parts. Note whether you have switches, wires, a circuit grid, and/or clips.

❷ Read the instructions for using the parts and study any Do's and Don'ts for building electronic circuits.

❸ Start with the first project listed in the kit (For Snap Circuits Jr. 100 this is the Electric Light and Switch). Follow the directions and assemble the project. Record your observations. Include a diagram of the finished circuit, label the parts, and describe how it works. Note whether the project worked as described.

Observations—First Electronics Project

❹ Once you have built the first project, assemble the second project (For Snap Circuits Jr. 100 this is the DC Motor and Switch). Follow the directions and assemble the project. Record your observations. Include a diagram of the finished circuit, label the parts, and describe how it works. Note whether the project worked as described.

Observations—Second Electronics Project

PHYSICS

Results

Now that you have a basic understanding of the kit and circuits, work through several more projects and record your observations for each.

Observations: _____ Project

Observations: _____ **Project**

PHYSICS

Observations: _____ **Project**

Observations: _____ **Project**

PHYSICS

III. Conclusions

A. Questions

❶ How easy or difficult was it to build a motor, light switch, alarm, or other device?

_____

_____

_____

_____

❷ What is a closed circuit?

_____

_____

_____

_____

❸ What is a DC motor?

_____

_____

_____

_____

❹ What is resistance and how does it work?

_____

_____

_____

_____

PHYSICS

❺ What is a fuse and what does it do?

_____

_____

_____

_____

B. Conclusions

Based on your observations, what conclusions can you draw from the circuits you built with the electronic circuit kit?

_____

_____

_____

_____

_____

_____

_____

_____

_____

_____

_____

PHYSICS

IV. Why?

In this experiment you built a variety of electric and integrated circuits by using an electronic circuit kit with an electrical grid. An electric circuit is a closed path through which electrons can flow. An integrated circuit is a combination of several electric circuits joined together along with components such as transistors, capacitors, and resistors. If you look inside many modern tools, toys, and appliances, you can find small circuit boards with sophisticated integrated circuits.

The invention of electric circuits led to profound changes in people's lives. The first battery that produced electrical current was made around 1800 by Alessandro Volta (1745-1827), an Italian physicist. In the US in the late 1800s Thomas Edison (1847-1931) experimented with electric circuits and designed the first coal-fired central power system that made it possible for people to have electric lighting at work and in their homes. The invention of many new appliances and machines was made possible by the availability of electricity and electric circuits. People no longer had to rely on candles or oil lamps for lighting or wood stoves for cooking food.

The addition of electronic components to electrical circuits to make integrated circuits has brought a huge variety of new technologies to the world. By knowing how electric circuits work, how to put them together, and how to make them convert electrical energy into light, sound, and motion, a growing number of tools are now available to scientists. Integrated circuits have made computers possible and are used in everything from programmable ovens to very advanced and complicated scientific instruments.

PHYSICS

V. Just For Fun

Use the parts from your kit to invent your own circuit. Try combining a motor or fan, sound, light, and switches. What other parts can you include? What can you create? Give your circuit a name and record your observations. Include a labeled diagram of your circuit and describe how it works.

Observations: _____

Experiment 11

Moving Marbles

Introduction

What happens when different size rolling marbles meet?

I. Think About It

❶ Do you think friction can affect a rolling marble? Why or why not?

_____

_____

_____

_____

_____

❷ Do you think a heavy marble could roll differently than a light marble? Why or why not?

_____

_____

_____

_____

❸ Do you think it might be harder to start a tennis ball rolling than a bowling ball? Why or why not?

_____

_____

_____

_____

PHYSICS

❹ Do you think inertia is involved when you hit a baseball with a bat? Why or why not?

_____

_____

_____

_____

_____

❺ Do you think it would be easier to catch a baseball that has more momentum or one that has less momentum? Why

_____

_____

_____

_____

_____

❻ What do you think would happen if there were no friction? Why?

_____

_____

_____

_____

_____

PHYSICS

II. Experiment 11: Moving Marbles   Date _____

Objective _____

_____

Hypothesis _____

_____

_____

Materials

several glass marbles of different sizes
several steel marbles of different sizes
cardboard tube, .7–1 meter long (2 1/2–3 ft)
scissors
black marking pen
ruler
letter scale or other small scale or balance

PHYSICS

EXPERIMENT

❶ Using the scale, weigh each of the marbles, both glass and steel. Keep track of how much each marble weighs by using the marking pen to label the marbles with numbers or letters, or you can note their colors. Record the information for each marble in *Part A* of the *Results* section.

❷ Take the cardboard tube and cut it in half lengthwise to make a trough. Measure the length of the cardboard trough, and mark the halfway point with the black marking pen.

❸ Beginning at the halfway mark, measure .3 meter (1 foot) in both directions, and put a mark at each of these measurements. This will give you one mark on each side of the halfway mark.

❹ The cardboard trough should now have three marks: one at the halfway point, and one on either side of the halfway mark, .3 meter (1 foot) away from it. The trough will be used as a track for the marbles.

❺ Take the marbles and, one by one, roll them down the trough. Notice how each one rolls (Does it roll straight? Is it easy to push off with your thumb? Does it pass the marks you drew?) In *Part B* of the *Results* section, describe how each marble rolls.

❻ Now place a glass marble on the center mark of the trough.

❼ Roll a glass marble of the same size toward the marble in the center. Watch the two marbles as they collide. Record your results in the *Results* section, *Part C.*

❽ Repeat Steps ❻ and ❼ with different size marbles. For example, try rolling a heavy marble toward a light marble and a light marble toward a heavy marble. Record your results in *Part D.*

Results

| Part A—Marble Descriptions ||
Marble	Weight

PHYSICS

Part B—How Marbles Roll

Part C—Same Size Colliding Marbles

Part D—Different Size Colliding Marbles

_____

_____

_____

_____

_____

_____

_____

_____

_____

PHYSICS

III. Conclusions

Review your observations. What conclusions can you draw from them?

_____

_____

_____

_____

_____

_____

_____

_____

_____

_____

_____

_____

_____

_____

_____

PHYSICS

IV. Why?

In this experiment you observed how the forces of *inertia, momentum,* and *friction* affected rolling and stationary marbles.

Inertia is the tendency of things to resist a change in motion. In physics there are two aspects to consider with regard to inertia: *mass* and *momentum*.

It is important to understand that *mass* and *weight* are different. Weight is a force. Mass is not. However, by weighing an object you can tell how much mass it has. The more mass an object has, the more it will weigh on Earth because gravity will exert more force on an object with a lot of mass than on an object with less mass. Without gravity, objects do not weigh anything. In space, where there is no gravity, a boulder would float in the same way a feather would. However, in space, the boulder and the feather would still have different masses. The boulder would still have more mass than the feather and, as a result, would still be harder to accelerate (speed up) than a feather—even in space.

The second aspect of inertia is the fact that an object with a lot of *momentum* is hard to stop. Momentum is inertia in motion; that is, mass that is moving. An object that has a large mass will have a large momentum. Also, an object moving at a fast rate of speed will have a large momentum. The mathematical equation for momentum is:

$$momentum = mass \times speed$$

Recall Newton's First Law of Motion that states: An object in motion will stay in motion unless acted on by an outside force, and an object at rest will stay at rest unless acted on by an outside force. Because inertia is the tendency of an object to resist a change in motion, objects that are stationary want to remain stationary, and objects that are moving want to remain in motion.

Although inertia keeps things moving, objects on Earth will eventually stop because a force acts on the object. This force is *friction*. Friction occurs when two objects rub against each other. Friction is the force that works in the direction opposite the direction of motion. Friction is what slows objects down and eventually causes them to stop. In the absence of friction, an object would keep moving forever and never stop. In this experiment, there is friction between the cardboard tube and the marbles.

PHYSICS

V. Just For Fun

Repeat the experiment, this time using a baseball, a basketball, and a golf ball. Think about whether you might have to modify the experimental setup. If you do want to modify it, what would you change and how would you change it?

Record your results. Draw conclusions about how differences in mass, inertia, momentum, speed, and friction affect this experiment.

Descriptions	
Object Name and Weight	**How It Rolls**

PHYSICS

Collision Descriptions

Conclusions

Review your observations. What conclusions can you draw about how differences in mass, inertia, momentum, speed, and friction affected this experiment?

_____

_____

_____

_____

_____

_____

_____

_____

_____

_____

_____

_____

_____

PHYSICS

Experiment 12

Accelerate to Win!

Introduction

How can knowing about velocity and acceleration help you win a race?

I. Think About It

❶ How do you think long distance runners win a race?

_____

_____

_____

_____

_____

❷ What do you think happens on the last lap of the Indy 500?

_____

_____

_____

_____

_____

❸ If you were riding in a horse race, what do you think you might do to win?

_____

_____

_____

_____

_____

❹ What do you think you would do if you were neck-and-neck with your best friend in a running race?

_____

_____

_____

_____

_____

❺ How do you think you could win a bike race with friends?

_____

_____

_____

_____

❻ How do you think you might train for winning a foot race? What do you think you would need to know?

_____

_____

_____

_____

_____

_____

_____

PHYSICS

II. Experiment 12: Accelerate to Win!   Date _____

Objective _____

_____

Hypothesis _____

_____

_____

Materials

stopwatch

compass

an open space large enough to run (park, schoolyard, playground, backyard, etc.)

5 markers of your choice to mark distances

a group of friends

EXPERIMENT

Imagine that you are training for the final race of an Olympic running race and you are determined to win. You have to go the full distance without stopping before the end and you need to go fast enough to win. You are going to follow your coach's recommendation and start slowly and then sprint as fast as you can for the last quarter of the race.

❶ Map out a straight "track" and mark a starting and stopping point.

❷ Take the compass and find out the direction you will be running in. Record this direction on the chart in the *Results* section.

❸ Measure the distance between the starting point and the stopping point by walking heel-to-toe and counting each step as one "foot." Record the distance here.

Length of track in "feet" _____

❹ Take your measurement of the length of the track and divide it into fourths. Record the distances to the points at which you will time your run. Each time point distance is measured from the previous time point.

Time point 1: d_1 (1/4 mark) _____

Time point 2: d_2 (1/2 mark) _____

Time point 3: d_3 (3/4 mark) _____

Time point 4: d_4 (Finish) _____

Now record distances d_1-d_4 on the chart in the Results section.

❺ On the track, measure with your feet the distance between each time point and mark time points d_1, d_2, and d_3. [d_4 (Finish) is already marked.]

❻ Pick one person to run the stopwatch. Have a second person use the chart in the *Results* section to record your time at each of the time points.

❼ Get ready! **Set! GO!**

❽ Repeat three or four times or until you are too tired to continue.

Results

❶ For each trial, use the formulas provided in the following chart to calculate the velocity at each time point. Space is provided for calculations. Record your results in the chart.

❷ For each trial, use the formulas provided in the chart to calculate the acceleration between each time point. Record your results in the chart.

PHYSICS

PHYSICS

Time Trial Results

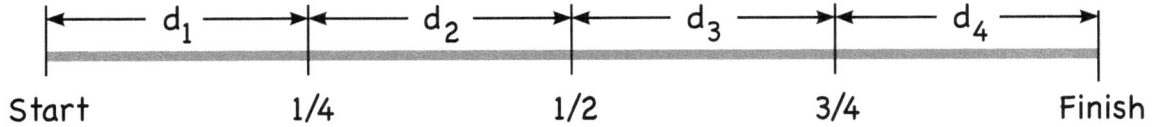

Start   1/4   1/2   3/4   Finish

Direction _____

Distance (in "feet")

d_1 _____ d_2 _____ d_3 _____ d_4 _____

Time (seconds) t_1 t_2 t_3 t_4

Trial 1 _____   _____   _____   _____

Trial 2 _____   _____   _____   _____

Trial 3 _____   _____   _____   _____

Velocity $v_1 = \dfrac{d_1}{t_1}$ $v_2 = \dfrac{d_2}{t_2}$ $v_3 = \dfrac{d_3}{t_3}$ $v_4 = \dfrac{d_4}{t_4}$

Trial 1 _____   _____   _____   _____

Trial 2 _____   _____   _____   _____

Trial 3 _____   _____   _____   _____

Acceleration* $a_1 = \left(\dfrac{v_2 - v_1}{|t_2 - t_1|}\right)$ $a_2 = \left(\dfrac{v_3 - v_2}{|t_3 - t_2|}\right)$ $a_3 = \left(\dfrac{v_4 - v_3}{|t_4 - t_3|}\right)$

Trial 1 _____   _____   _____

Trial 2 _____   _____   _____

Trial 3 _____   _____   _____

*Note: For acceleration, time is always a positive number. In the acceleration formula, the change in time (Δt) is written as $|t_f - t_i|$ to show that the result is expressed as a positive number.

A Place for Calculations

III. Conclusions

A. Questions

❶ Which segment did you run with the fastest velocity? Why?

_____

_____

_____

❷ Which segment did you run with the slowest velocity? Why?

_____

_____

_____

❸ What can you notice about your acceleration in the different trials?

_____

_____

_____

❹ In how many segments was your acceleration positive? negative? Which ones? Why?

_____

_____

_____

_____

B. Conclusions

Compare your trials. How was your performance in each? Were you faster or slower on the third trial? Explain your observations and results.

_____

_____

_____

_____

_____

_____

_____

_____

_____

_____

_____

_____

_____

_____

_____

PHYSICS

IV. Why?

By measuring the time it takes you to run between different points of known distance, you can calculate your velocity and acceleration. If you were training for the Olympics, by knowing how much energy you have and how fast you can go for how long, you could monitor how well you are doing in each run. You might notice that if you start out the run with a fast pace and accelerate too much near the beginning of the race, you are likely to run out of energy and slow down, decelerating near the finish line. Running out of energy before the end of the race won't help you to win, but you can learn how to start more slowly, run at a steady pace, and then accelerate at the finish.

V. Just For Fun

Run every day for a few weeks, recording the date and length of time. Then repeat the experiment. Record your results in the following chart. Have your times improved? What changes have you made in how you run a race? Use additional paper for observations such as the route you follow, weather, etc.

PHYSICS

Date and Length of Time for Each Run

Date	Time	Date	Time	Date	Time

Time Trial Results

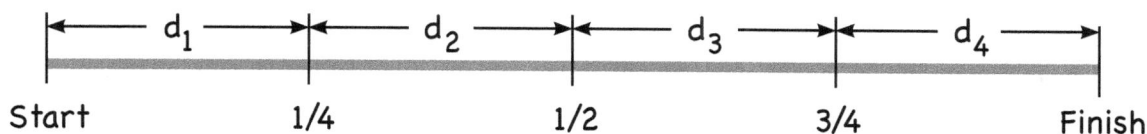

$$\xleftarrow{\hspace{1em}} d_1 \xrightarrow{\hspace{1em}} \xleftarrow{\hspace{1em}} d_2 \xrightarrow{\hspace{1em}} \xleftarrow{\hspace{1em}} d_3 \xrightarrow{\hspace{1em}} \xleftarrow{\hspace{1em}} d_4 \xrightarrow{\hspace{1em}}$$

Start   1/4   1/2   3/4   Finish

Direction _____

Distance (in "feet")

d_1 _____ d_2 _____ d_3 _____ d_4 _____

Time (seconds) t_1 t_2 t_3 t_4

Trial 1 _____	_____	_____	_____
Trial 2 _____	_____	_____	_____
Trial 3 _____	_____	_____	_____

Velocity $v_1 = \dfrac{d_1}{t_1}$ $v_2 = \dfrac{d_2}{t_2}$ $v_3 = \dfrac{d_3}{t_3}$ $v_4 = \dfrac{d_4}{t_4}$

Trial 1 _____	_____	_____	_____
Trial 2 _____	_____	_____	_____
Trial 3 _____	_____	_____	_____

Acceleration* $a_1 = \left(\dfrac{v_2 - v_1}{|t_2 - t_1|} \right)$ $a_2 = \left(\dfrac{v_3 - v_2}{|t_3 - t_2|} \right)$ $a_3 = \left(\dfrac{v_4 - v_3}{|t_4 - t_3|} \right)$

Trial 1 _____	_____	_____
Trial 2 _____	_____	_____
Trial 3 _____	_____	_____

*Note: For acceleration, time is always a positive number. In the acceleration formula, the change in time (Δt) is written as $|t_f - t_i|$ to show that the result is expressed as a positive number.

There is space for calculations on the following page.

A Place for Your Calculations

PHYSICS

Experiment 13

Around and Around

Introduction

Measure tangential speed!

I. Think About It

❶ If you run around your backyard in a circle, how far do you go?

_____

_____

❷ If you run around a circular running track how far do you go?

_____

_____

❸ If you could run around the Earth, how far would you go?

_____

_____

❹ Which is bigger, the diameter of a circle you can run in your backyard or the diameter of the Earth?

_____

_____

PHYSICS

❺ If you could complete one lap around the Earth in the same length of time that you could complete one lap around your backyard, would you be running faster or slower in your trip around the Earth? Why?

_____

_____

_____

_____

_____

❻ Do you think tangential speed is the same at any location on a disk that is spinning? Why or why not?

_____

_____

_____

_____

_____

❼ Do you think tangential speed and rotational speed are related? Why or why not?

_____

_____

_____

_____

_____

PHYSICS

II. Experiment 13: Around and Around   Date _____

Objective _____

_____

_____

Hypothesis _____

_____

Materials

pen or pencil
marking pen
thumbtack or pushpin
3 pieces of string—approximate sizes:
    10 cm [4 in.]; 15 cm [6 in.]; 20 cm [8 in.]
tape
ruler
large piece of white paper (bigger than 30 cm [12 in.] square)

EXPERIMENT

❶ Lay the white sheet of paper on a flat, firm surface and use the thumbtack or pushpin to pin one end of the shortest string to the center of the paper.

❷ Measure the string beginning at the thumbtack and put a mark at 5 cm (2 in.).

❸ Take the pen or pencil and place it at the 5 cm mark on the string. Then wrap the extra length of string around the pen or pencil and fasten it with tape.

❹ Holding the thumbtack down, move the pen away from the thumbtack until the string is stretched out.

❺ Place the point of the pen in contact with the paper, holding the pen in a perpendicular position. Draw a circle around the center point.

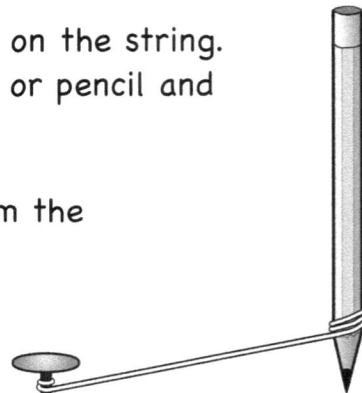

⑥ Repeat Steps ❶-❺ with the other two pieces of string. For the middle size string the pen will be 10 cm (4 in.) from the thumbtack, and for the longest piece of string the pen will be 15 cm (6 in.) from the thumbtack.

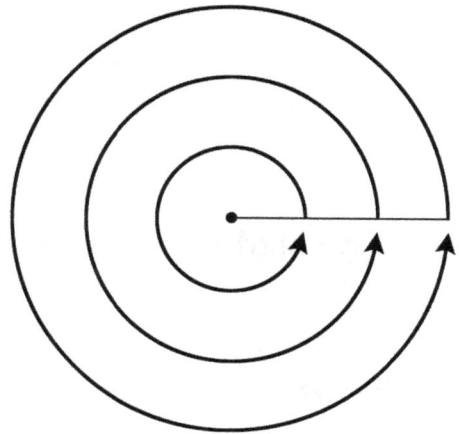

Results

Using the formulas provided in the table on the next page, calculate the tangential speed of each circle.

❶ With the ruler, measure the radius of each circle and write these numbers in the table.

❷ Using the equation in the calculation box, calculate the circumference for each circle.

❸ Calculate the tangential speed for each circle, assuming a rotational speed of 1 RPM (revolution per minute).

Calculating Tangential Speed

	①	②	③
String Length	5 cm	10 cm	15 cm

Circle:

Radius _____   _____   _____

Calculate the Circumference [π (pi) = 3.14]

①	②	③
c = 2πr c = 2πr · _____ c = _____	c = 2πr c = 2πr · _____ c = _____	c = 2πr c = 2πr · _____ c = _____

Circumference _____   _____   _____

Calculate the tangential speed for one revolution (1 RPM)

Tangential speed = distance traveled/time

> Note: Distance traveled is one revolution—the circumference of the circle (c).

> Time (t) equals one minute for this problem.

Tangential speed (S_T) = c/t

Calculation   ①   ②   ③

①	②	③
$S_T = c/t$ $S_T =$ _____ /t	$S_T = c/t$ $S_T =$ _____ /t	$S_T = c/t$ $S_T =$ _____ /t

Tangential Speed _____   _____   _____

III. Conclusions

Which circle has the highest tangential speed? _____

Which circle has the lowest tangential speed? _____

What conclusions can you draw from your observations?

_____

_____

_____

_____

_____

_____

_____

_____

_____

_____

_____

_____

_____

_____

IV. Why?

One of the great things about physics is that mathematics will often confirm your experience. If you hit the gas pedal in a car, you will feel the acceleration, and you can show this with mathematics. If you suddenly stop by hitting the brake, you will feel the change in momentum, and you can also show this mathematically.

By doing this experiment you can experience how tangential speeds are greater on the outside of a circle than towards the center. Another way to make this observation is to put your finger on a rotating disk and note how your finger moves faster when it is near the outside of the disk than when it is near the center. If you sit first on the outside of a merry-go-round that is spinning and then sit near the center, you can feel the difference in tangential speed.

In this experiment you calculated the difference in tangential speed between three circles of different sizes. The larger circles have a greater radius than the smaller circles, and thus they have a larger circumference, or distance around the circle. If the rotational speed is the same for all three circles, you can see that the tangential speed increases as the size of the circle increases. This is what you experience if you sit at different distances from the center of a merry-go-round.

So the next time you go to the amusement park, if you want to spin faster or spin slower on the Spinning Top Ride, you'll know whether to choose a spot closer to or farther away from the center!

V. Just For Fun

What objects in motion have you observed that have tangential speed? List them on the following page.

How would you measure the tangential speed of one of these objects? Record your ideas about measuring the tangential speed of this object.

Tangential Speed

Moving objects that have tangential speed

Ideas for measuring tangential speed of _____

Experiment 14

Hidden Treasure

Introduction

Make your own treasure map!

I. Think About It

❶ If you are hiking in a wilderness area where a GPS device doesn't work, what tools could you use to find your way? Why?

_____

_____

_____

_____

_____

❷ What facts do you think you can learn about an area by using a map?

_____

_____

_____

_____

_____

❸ If you were making a map of the area where you live, what details would you include? Why? How do you think someone else might use this map?

_____

_____

_____

_____

GEOLOGY

❹   What do you think you could discover if you had a topographic map, a compass, a rock hammer, and a rock and mineral test kit in your backpack?

_____

_____

_____

_____

_____

❺   What advantages do you think you might have if you added a GPS device to your backpack? Why?

_____

_____

_____

_____

_____

GEOLOGY

❻   What other electronic tools do you think you might use to help you explore the geology of an area? What do you think you might discover by using them?

_____

_____

_____

_____

_____

II. Experiment 14: Hidden Treasure

Date _____

Objective _____

_____

Hypothesis _____

_____

_____

Materials

> pencil, pen, colored pencils
> compass
> a small jar or container with a lid
> small items to place in the jar (treasure)
> garden trowel (if needed)

EXPERIMENT

❶ Find some small objects to be your treasure and put them in the jar.

❷ Select an area near your home to make a map of. This area can be your front or back yard, a park, or other open space.

❸ Using each of your feet as a one-foot ruler, measure the outline of the area, walking heel-to-toe around it. Count your steps and notice if your path goes in a straight line or curves around objects.

After measuring a side of the area, draw that side in the space provided on the next page. Make your map as accurate as possible, noting on your map now many steps (feet) there are on each side of the area.

❹ Once you have finished measuring and drawing the outline of the area, hold the compass and turn around until the needle lines up with the north (N) symbol. Note this direction (north) on your map, drawing an arrow and an "N." Next, note south (S), east (E), and west (W), again drawing arrows and using the letters.

Treasure Map

GEOLOGY

❺ Add details to your map. Measure the distance to trees, shrubs, buildings, or other features by walking heel-to-toe and counting your steps. On the map, record the direction, the distance, and a drawing of the object. If you need to find out where to locate an object, such as a tree, on your map, pick a starting point (for example, the object in the corner of your map) and measure the distance as you walk toward the tree you want to position on the map.

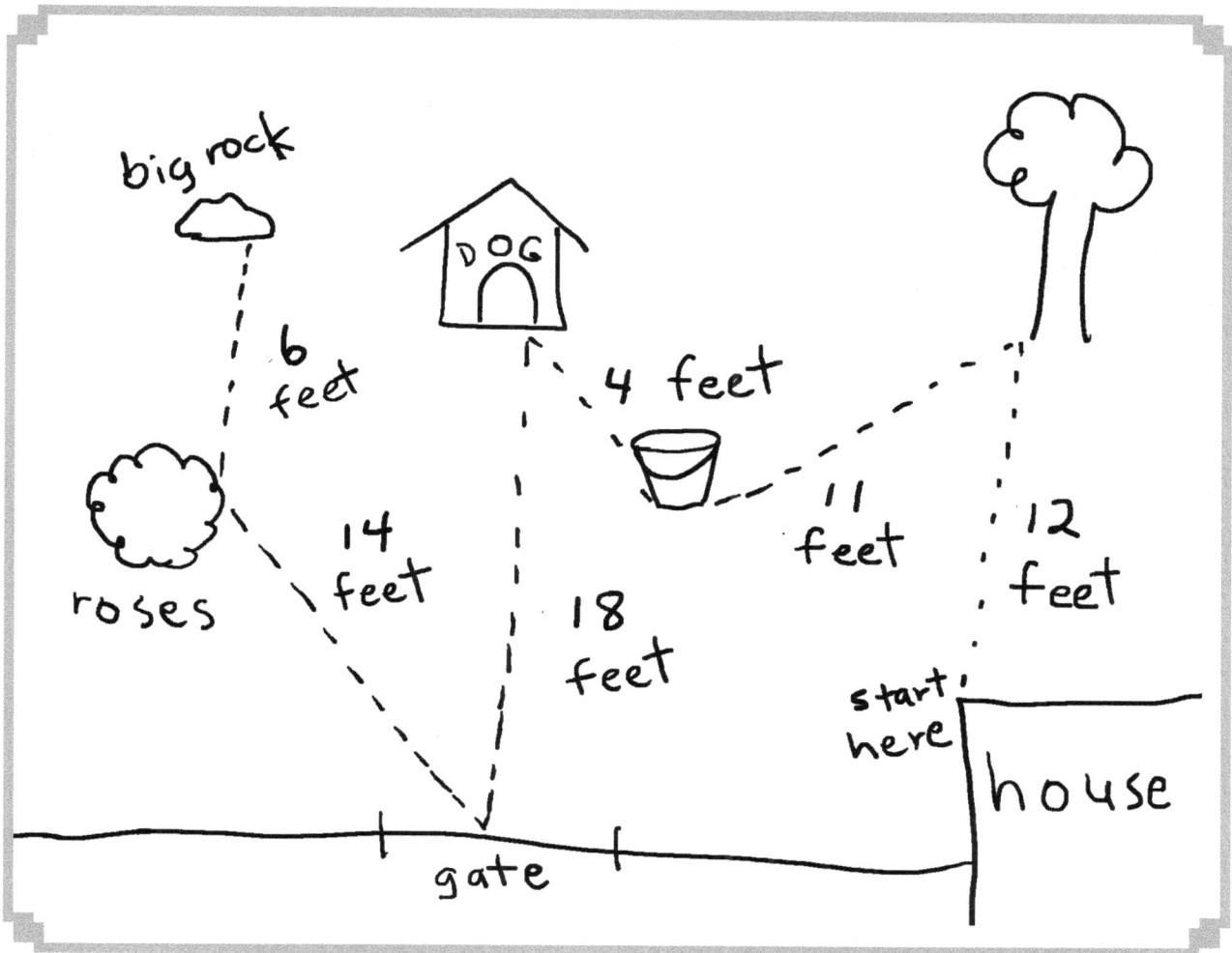

❻ Now pick a location in the area you have mapped and bury or hide your treasure there. Record the location of the treasure on your map.

❼ Give your map to a friend and see if they can use it to find your hidden treasure.

Results

❶ Observe how easy or difficult it is for your friend to find your hidden treasure.

❷ In the following chart, record the number of attempts it takes your friend to find the hidden treasure. Also record how much help you have to give them before they find the treasure. Note any adjustments you make to your map to make it more accurate.

Treasure Map Results

Number of Attempts	
Type of Help	
Adjustments to Map	

GEOLOGY

III. Conclusions

What conclusions can you draw from your observations?

_____

_____

_____

_____

_____

_____

_____

_____

_____

_____

_____

_____

_____

_____

_____

GEOLOGY

IV. Why?

As with any science, geology has been transformed as a result of technological advances. Modern tools allow geologists to study aspects of Earth's features and dynamics that were unavailable to early geologists.

Hand tools are easy to carry and easy to use, even if they are not necessarily technologically advanced. The rock hammer, map, compass, and rock and mineral test kit are tools that fit nicely in a backpack, are generally inexpensive, and are easy to use.

Modern geologists use a variety of electronic tools to study Earth. Although electronic tools are dependent on a variety of other technologies working properly (e.g., batteries, satellites, motors), they can offer substantial advantages over older or non-technological tools. For example, a global positioning system may find your location more accurately than you can find it on a paper map, and paper maps can become out of date. Some GPS devices can record the route you have walked and make it easier for you to find your way back than it would be if you were using a compass and map. However, GPS devices don't work everywhere and batteries may fail, so a geologist out in the field may also do well to know how to use a paper map and a compass.

Today's geologists also have ground penetrating radar available that can image below the surface of the Earth without disrupting the ground. GPR can be used from satellites as well as from the surface of the Earth. Satellite GPR and other remote sensing devices can help geologists find out about things such as the below surface structure of landforms and where to find groundwater, even in locations that are difficult to access. Archaeologists are now using satellite GPR to find ancient ruins that have been buried over the years.

Other tools used by geologists include drills and rock and mineral test kits. These tools help geologists directly test samples taken from the Earth to find out more about the composition and structure of landforms in different areas. Seismometers and seismographs help geologists understand what makes earthquakes and volcanoes occur and what the layers of the Earth may be like.

As technology advances, geologists will be able to find out more and more about Earth's structure and dynamics.

GEOLOGY

V. Just For Fun

Review the results of the experiment and evaluate your map. When your friend looked for the treasure, what parts of the map worked and what parts didn't work? Think of as many ways as you can to make your map better. Indicate a new location to hide the treasure. Now have a friend use the map to find the treasure in its new location.

Evaluate your revised map and record your results.

Treasure Map Results #2	
Number of Attempts	
Type of Help	
How well did the revised map work compared to the original map? Why?	

Experiment 15

Using Satellite Images

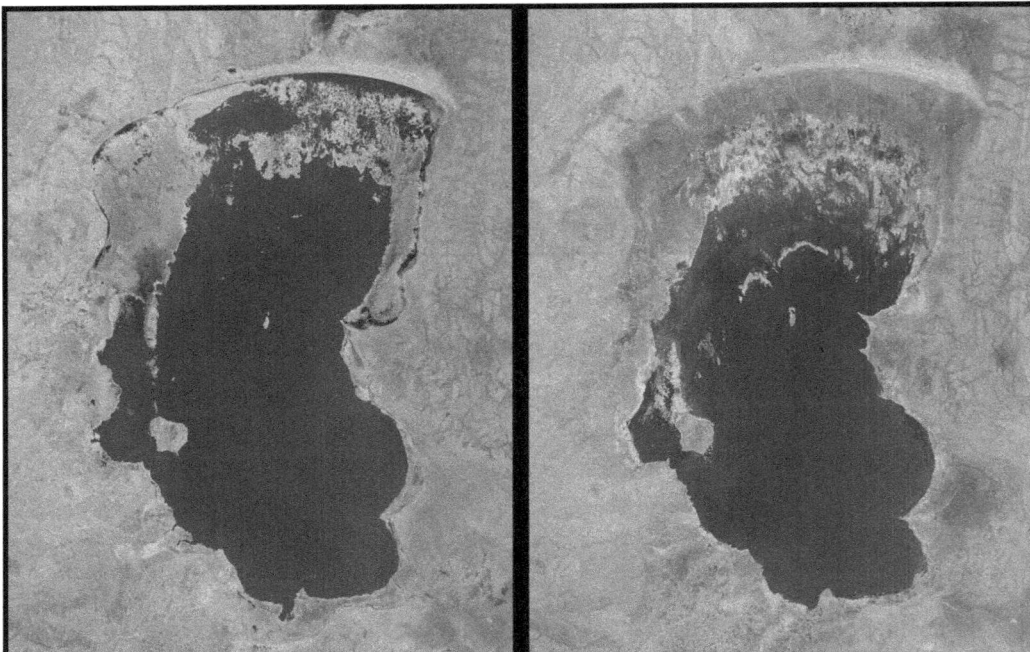

Two satellite images of Lake Chilwa, Malawi show it is shrinking

USGS Landsat photos

Introduction

Explore using satellite imagery to study Earth's spheres.

I. Think About It

❶ List a few events that you think could change the geosphere.

_____

_____

_____

_____

_____

❷ List a few events that you think could change the atmosphere.

_____

_____

_____

_____

_____

❸ List a few events that you think could change the biosphere.

_____

_____

_____

_____

_____

GEOLOGY

❹ List a few events that you think could change the hydrosphere.

_____

_____

_____

_____

_____

❺ Do you think changes to the geosphere can occur below the Earth's crust? Why or why not?

_____

_____

_____

_____

❻ Do you think there are events that happen in space that can affect any of Earth's spheres? Why or why not?

_____

_____

_____

_____

GEOLOGY

II. Experiment 15: Using Satellite Images   Date _____

Objective   _____

_____

Materials

computer with internet access
printer and paper (optional)
colored pencils (optional)

EXPERIMENT

The US Geological Survey (USGS) operates a satellite called Landsat that photographs Earth's surface. Their online Land Remote Sensing Image Gallery shows sets of images that compare changes to Earth over time.

❶ Go to the US Geological Survey Landsat website:

https://remotesensing.usgs.gov/gallery/

Spend some time looking through the collections of images that show changes to Earth's surface.

❷ Select one set of images that shows a change to the geosphere. In the *Download Image* section, click on the small size image file and download it. If you get a "zip file" that contains the images, you will need a program that unzips files for you to be able to look at the images. Make a folder on your computer to keep the image files in.

If possible, print the images, label them, and insert them in your *Laboratory Notebook*. If you can't print them, you can refer to the files on your computer as needed.

Record your observations in the *Results* section.

❸ Select one set of images that shows a change to the atmosphere. Download and print them if possible. Record your observations in the *Results* section.

❹ Select one set of images that shows a change to the hydrosphere. Download and print them if possible. Record your observations in the *Results* section.

GEOLOGY

❺ Select one set of images that shows a change to the biosphere. Download and print them if possible. Record your observations in the *Results* section.

Results

A Change in the Geosphere

Location:

Period of time covered:

Description of changes to the geosphere. Include a rough sketch.

How do you think this information could be used to help the environment and/or to help people?

GEOLOGY

A Change in the Atmosphere

Location:

Period of time covered:

Description of changes to the atmosphere. Include a rough sketch.

How do you think this information could be used to help the environment and/or to help people?

A Change in the Hydrosphere

Location:

Period of time covered:

Description of changes to the hydrosphere. Include a rough sketch.

How do you think this information could be used to help the environment and/or to help people?

GEOLOGY

A Change in the Biosphere

Location:

Period of time covered:

Description of changes to the biosphere. Include a rough sketch.

How do you think this information could be used to help the environment and/or to help people?

III. Conclusions

Discuss what you learned by observing images taken of Earth from space. How do you think satellites have changed what we can learn about Earth?

_____

_____

_____

_____

_____

_____

_____

_____

_____

_____

_____

_____

_____

_____

_____

GEOLOGY

IV. Why?

One way scientists observe changes in Earth's spheres is through satellite images. By using images from space, scientists can watch natural events such as hurricanes, typhoons, dust storms, volcanoes, and wildfires and how they affect the different spheres of Earth. Changes in forests, deserts, glaciers, rivers, icebergs, oceans, pollution, and population can be observed over time to learn how natural and human activity affect the atmosphere, hydrosphere, biosphere, geosphere, and magnetosphere.

By examining images taken of the same location over the course of several days, weeks, months, or years, scientists studying different parts of Earth's spheres can learn how changes to Earth's interconnected system occur over time. For example, if a particular typhoon were followed for several days, observations could be made about where and how it forms and grows, what route it follows, and where and when it breaks apart. This information could help scientists predict when future storms might occur. The effects of the passage of the typhoon could also be studied, showing how the biosphere, geosphere, hydrosphere, and atmosphere were affected.

Using satellite images to study different natural events allows scientists to gather information that can be used to help people be better prepared for storms and other natural events and help them find ways to protect their communities and surrounding environments. The effects of population growth can be seen as cities and towns expand, and satellite images may be used to help find solutions to the impact on the environment. Air and water pollution can be observed, helping solutions to be found, and the effects of deforestation and other changes in plant distribution can be studied.

In addition to photographs, satellites use an array of other technologies, such as radar, laser beams, and thermal imaging, to further record details of Earth's system. Putting together information from these different types of imagery gives scientists a more complete picture of Earth and its spheres.

GEOLOGY

V. Just For Fun

Discover more about Earth's spheres through satellite images.

NASA's Earth Observatory website has photographs and videos taken from the International Space Station.

http://earthobservatory.nasa.gov/

On the top menu bar click on *Images* to find collections by topic. Select *Natural Hazards* and explore satellite images of several natural disasters.

Go back to the Home pages and scroll down to find the *Special Collections* groups of photographs to explore.

NASA's Gateway to Astronaut Photography of Earth

http://eol.jsc.nasa.gov/Collections/EarthFromSpace/

Select a topic from the left menu bar. On the next screen select what you would like to view and click *Start Search*. Play around to see what else you can find on this site. Record some of your observations.

GEOLOGY

Observations of Earth from Space

What did you see that was the most surprising?

Observations of Earth from Space

What did you see that was the most beautiful?

What would you like to explore further?

What could you notice that you wouldn't be able to see from the ground?

If you were a geologist, how do you think you would use some of the satellite images you saw?

GEOLOGY

Experiment 16

Modeling Earth's Layers

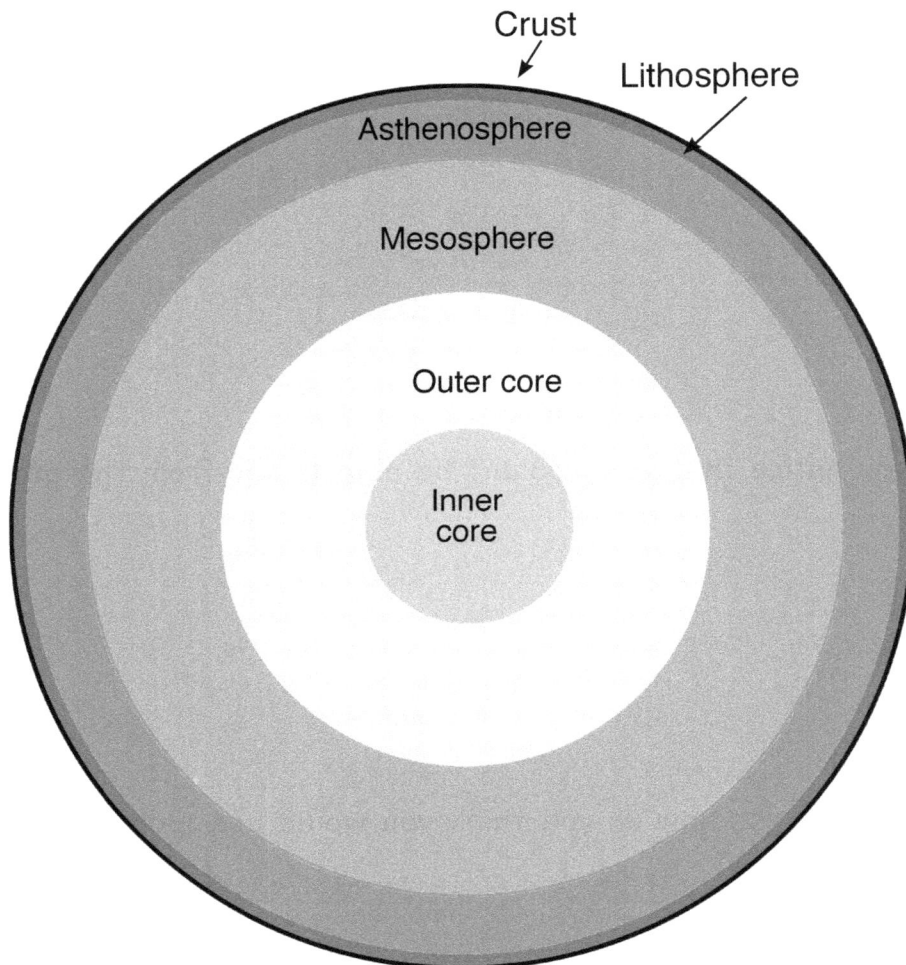

Introduction

Learn more about the geosphere by making a model.

I. Think About It

❶ What layers do you think Earth has and how are they arranged?

_____

_____

_____

_____

_____

❷ How do we know how many layers are in the geosphere and what they are most likely made of?

_____

_____

_____

_____

_____

❸ What do you think the boundaries between Earth's layers look like?

_____

_____

_____

_____

_____

❹ What do you think Earth would be like if its layers were arranged in a different order?

_____

_____

_____

_____

_____

❺ What do you think scientists might find out about Earth's different layers if they could drill down to the center of the Earth?

_____

_____

_____

_____

_____

❻ What types of materials do you think you could use to represent Earth's different layers in a model?

_____

_____

_____

_____

_____

GEOLOGY

II. Experiment 16: Modeling Earth's Layers   Date _____

Objective _____

_____

Hypothesis _____

_____

_____

Materials

Some suggestions for model making materials:
modeling clay of different colors
marble or steel ball
ingredients to make various colored cakes
materials for making paper mache
Styrofoam balls

EXPERIMENT

❶ In the space on the next page, list the names of Earth's layers and a summary of known features (hard, brittle, soft, hot, molten etc.)

Layer	Features

❷ Imagine it is your job to create a model that resembles Earth's layers as closely as possible. Spend a few minutes thinking about how you might accomplish this task, what materials you might use, and how easy or difficult it would be to create a model of Earth. Record your ideas.

Ideas for Modeling Earth's Layers

❸ Review your notes and build a model of Earth's layers based on your own ideas.

Results

❶ Draw a diagram of your model. Label Earth's layers and the materials you used to make them.

GEOLOGY

❷ Think about how easy or difficult it was to create the model of Earth and then write about what you considered important to include and what you decided to ignore and why. For example, you cannot build an exact model of Earth's size, so "size" is something you could scale or ignore. Or you might have thought that using gelatin or peanut butter would be a great idea for modeling the asthenosphere, but creating a spherical layer might be challenging. Instead, you might have decided to model the layers on a flat surface, ignoring the spherical feature of the layers and focusing on their other features.

Notes About Model Making

III. Conclusions

Discuss what you learned by creating a model of Earth and Earth's layers.

_____

_____

_____

_____

_____

_____

_____

_____

_____

_____

_____

_____

_____

_____

GEOLOGY

IV. Why?

Model building is an important aspect of scientific investigation. Scientists build models of their ideas to help them see how things work and how to think of new ways to understand scientific phenomena. Scientists can use models to understand how Earth's layers interact, how deep they are, and what rocks and minerals they might contain.

However, model building isn't always easy and finding the right materials to model important features can be a challenge. For example, you might have elected to use gelatin or peanut butter to model the soft layer of the asthenosphere but found it challenging to create a spherical layer. Or you might have decided to use modeling clay for all the layers making it easy to create spheres but ignoring the fact that the layers have different textures.

Models don't generally duplicate all the important features of the idea or object a scientist is trying to understand. Knowing which features to include and which to ignore depends on what the model is being used to explore.

V. Just For Fun

Think about how you might create P and/or S waves through a model of Earth's layers. Would you have to use different materials than you did for the model you just created? Write down your ideas.

Now use your ideas to make a flat model of one or more of Earth's layers and see if you can get a wave to pass through. Record your results. Include what type of wave you think you created.

GEOLOGY

Making Waves

Ideas

Model

GEOLOGY

Experiment 17

Exploring Cloud Formation

Courtesy of US Fish and Wildlife Service, photo by Steve Hillebrand

Introduction

Do you think you can make clouds in a bottle?

I. Think About It

❶ How do you think clouds are formed?

_____

_____

_____

_____

_____

❷ How many different kinds of clouds have you observed? How would you describe them?

_____

_____

_____

_____

_____

❸ What factors in the atmosphere do you think affect cloud formation?

_____

_____

_____

_____

_____

GEOLOGY

❹ Do you think clouds are more likely to form over an ocean or a desert? Why?

_____

_____

_____

_____

❺ What do you think would happen to life on Earth if there were no clouds? Why?

_____

_____

_____

_____

❻ How do you think clouds move? Do you think they go far? Do you think they go fast? Why or why not?

_____

_____

_____

_____

GEOLOGY

II. Experiment 17: Exploring Cloud Formation   Date _____

Objective _____

_____

Hypothesis _____

_____

_____

Materials

2 liter (2 quart) plastic bottle with cap
warm water
matches

EXPERIMENT

❶ Pour warm water into the plastic bottle until it is about 1/4 full. Put the cap on the bottle.

❷ Light a match and remove the cap from the bottle. Drop the match in the bottle and quickly replace the cap.

❸ Squeeze the plastic bottle near the bottom and release. Notice what happens to the air in the bottle as you do this.

❹ Record your observations in the chart in the *Results* section.

❺ Repeat this experiment, filling the plastic bottle 1/2 full, 2/3 full, and then almost full. After each experiment, empty the bottle and start with fresh warm water. Record your observations each time.

GEOLOGY

Results

Cloud Formation Observations

Water Level in Bottle	Observations
1/4 Full	
1/2 Full	
2/3 Full	
Almost Full	

III. Conclusions

What conclusions can you draw from your observations? How would you relate your results to what happens in the atmosphere?

_____

_____

_____

_____

_____

_____

_____

_____

_____

_____

_____

_____

_____

GEOLOGY

IV. Why?

Water vapor is essential for life. It is the most important gas for keeping Earth warm and is also part of Earth's water cycle. Liquid water evaporates from bodies of water and from the land and enters the atmosphere as water vapor. The water vapor later condenses, turning back to the liquid state, forming clouds and then falling to Earth as precipitation.

Since warm air can hold more water vapor than cold air, cooling the air to the point where it can no longer hold all the water vapor will cause the water vapor to condense and turn to the liquid state. The temperature at which condensation occurs is called the dew point. The dew point varies depending on the amount of water vapor in the air (humidity).

The most common way for air to be cooled is through lifting. Air moves into an area of the atmosphere that is at a lower pressure, causing the air to expand. Energy is required for this expansion, taking heat away from the air and cooling it. As the air cools, some of the water vapor will condense around dust particles in the air, forming water droplets.

The reverse happens as air sinks. As it encounters higher pressures at lower altitudes, the increased pressure squeezes the air, adding heat and allowing the air to once again hold more water vapor. This can cause clouds to evaporate as the liquid water turns to vapor.

In order for water droplets to fall as precipitation, they need to become larger and heavier. One process by which this happens is called the collision and coalescence process or the warm rain process. During this process, water droplets in clouds collide and stick together (coalesce) to form larger drops.

Since warm air can hold more water vapor than cold air and heat makes water evaporate, the highest levels of water vapor in the atmosphere are over the oceans in the equatorial region where the heat evaporates water from the oceans and more of this water vapor can be held in the warm air. The lowest levels of water vapor are found over the dry deserts where there is little water to evaporate and at the poles where the air is cold and the water is tied up as ice.

V. Just For Fun

Observe the clouds in your area for two weeks or more and record your daily weather observations. Describe the weather conditions and describe and sketch the clouds that you see. Follow the format below to continue your chart on separate paper.

Check online daily for humidity, dew point, and low and high temperatures. To find this information you can do a simple search such as "dew point today" with your location. Record this information on your chart.

At the end of your observation period, look at the data on your chart. Is there a relationship between the types of clouds and the weather? Do you think the types of clouds are related to the temperature, dew point, and/or humidity? Record your conclusions and fasten the observation chart pages in the *Laboratory Notebook*.

Cloud Formation Observations

	Observations (Weather, Types of Clouds)
Date: _____	
Humidity: _____	
Dew point: _____	
High Temp: _____	
Low Temp: _____	

GEOLOGY

Experiment 18

Measuring Distances

Mars Rover   Image Credit: NASA/JPL/Cornell University

Introduction

Use a simple triangulation method to measure the distance of a faraway object.

I. Think About It

❶ Do you think ancient astronomers knew the distance to different stars? Why or why not?

_____

_____

_____

_____

❷ Do you think simple tools could be used to measure the distance to faraway objects? Why or why not?

_____

_____

_____

_____

❸ Do you think the development of advanced technology has helped today's astronomers find out the distance to different stars? Why or why not?

_____

_____

_____

_____

ASTRONOMY

❹ Do you think you could measure the distance to a faraway object without using mathematical formulas? Why or why not?

_____

_____

_____

_____

_____

❺ Do you think a knowledge of mathematics is important to doing astronomy? Why or why not?

_____

_____

_____

_____

_____

❻ Do you think the invention of space telescopes has helped astronomers find out the distance to different stars? Why or why not?

_____

_____

_____

_____

_____

ASTRONOMY

II. Experiment 18: Measuring Distances   Date _____

Objective _____

_____

Hypothesis _____

_____

_____

Materials

two sticks (used for marking)
two rulers
tape
string, several meters long (several yards)
protractor
square grid or graph paper (included in this chapter)

EXPERIMENT

❶ Find a wide open space with a distant object. The space can be a field, a city street, or even your own backyard.

❷ Pick two observation points, and place the sticks at these points. Mark one observation point "A" and the other "B."

❸ Take two rulers and tape them together at one end, making a right angle.

❹ Place the corner of the double ruler on observation point "A" with one end pointing towards the object you want to measure and the other end pointing towards observation point "B."

❺ Attach the string to the stick at observation point "A," and stretch it out along the side of the double ruler pointing towards observation point "B." The string will be used as a guide so that you walk in a straight line.

❻ Holding the string, walk heel-to-toe from observation point "A" to observation point "B," making sure the string is still pointing at a 90 degree angle in the direction of point "B." Count your steps. Each step will equal one "foot." (Note: You may need to adjust the location of point "B" to maintain the right angle at point "A.")

❼ When you get to point "B," attach the string to the stick. Check to make sure the string is still pointing in the same direction as the ruler. In the *Results* section, record the number of steps between point "A" and point "B."

❽ From observation point "B" find the object whose distance you want to measure. Place the protractor on the string so that you can measure the angle between point "B" and the distant object.

❾ Record the angle between point "B" and the distant object.

Distant Object

Experimental Setup

90°

string

angle x

steps

protractor

Point A

Point B

ASTRONOMY

Results

Now that you have collected your data, use grid paper and a modeling technique to measure the distance to the object.

❶ On the grid paper on the following page, mark point "A."

❷ Using one square for each step, mark point "B" on the grid.

❸ Draw a line from point "A" to point "B." This is line "AB."

❹ Draw a line from point "A" towards the distant object. This line should be at a 90 degree angle to line AB. Label this line "y."

❺ Using your protractor, at point "B" on your graph mark the angle you measured at point "B." Draw a line from point "B" to the distant object using this angle. Extend this line until it intersects with line "y." (You may have to extend line "y" in order for the two lines to intersect.)

❻ Count the number of squares from point "A" along line "y" to the distant object.

❼ Assuming that each of your steps is one foot, how far away is the distant object? Record your answers below.

Number of steps—Point A to Point B _____

Angle at Point B _____

Number of squares—Point A to distant object _____

Distance of object in feet _____

ASTRONOMY

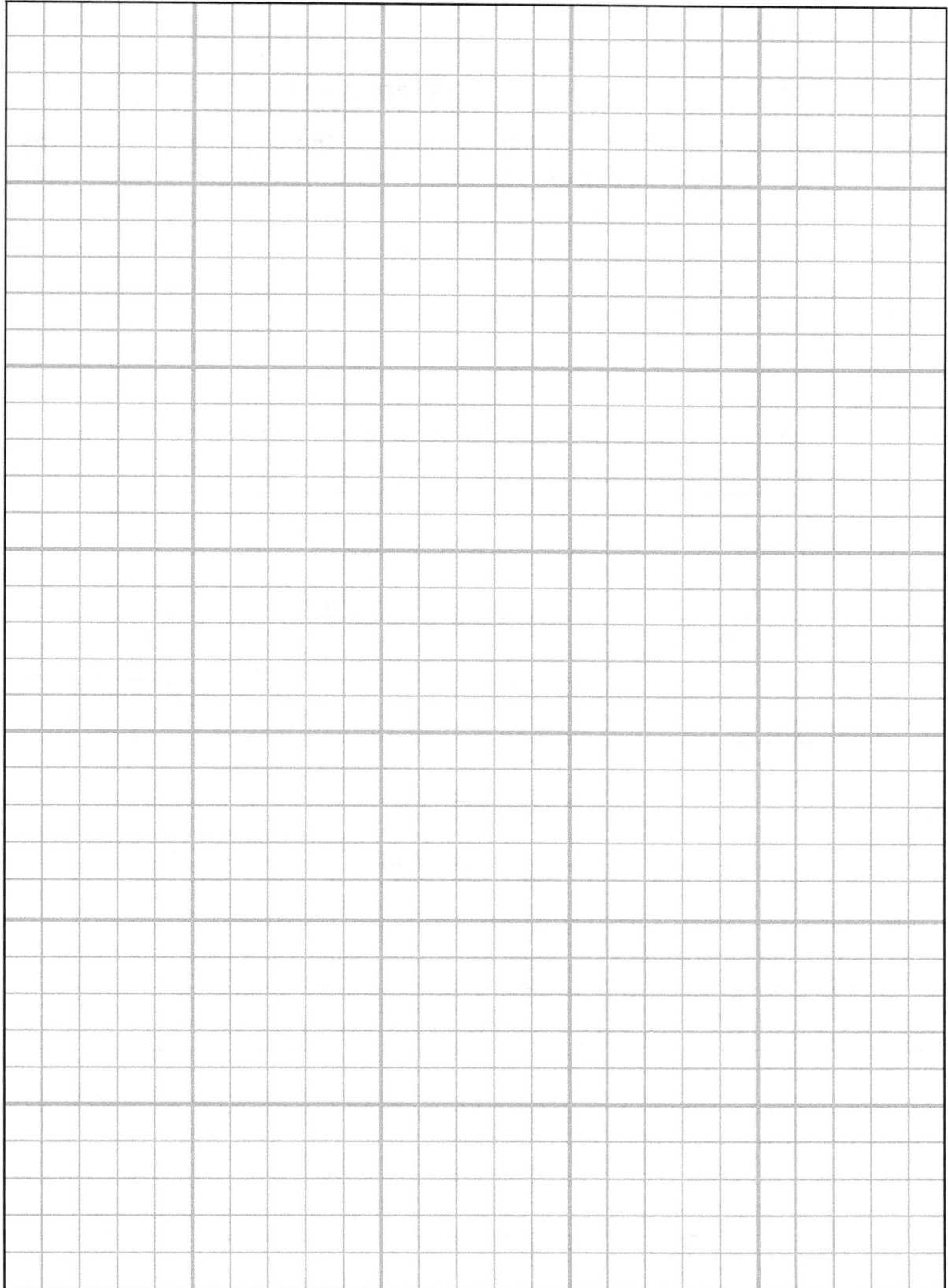

III. Conclusion

What conclusions can you draw from your observations?

Summarize how easy or difficult it was to measure the distance of a faraway object. Write down any problems or sources of error you might have noticed.

_____

_____

_____

_____

_____

_____

_____

_____

_____

_____

_____

_____

_____

_____

ASTRONOMY

IV. Why?

In this experiment you were able to calculate the distance of a faraway object by using two points, some basic geometry, graph paper, and a method called *triangulation*. Triangulation uses the concept of *similar triangles* to estimate distances. In geometry, similar triangles are those that have exactly the same shape and only differ in size. Because the three angles in a triangle always add up to 180 degrees, if we know the sizes of two of the angles, we can add them together and subtract the total from 180 degrees to get the size of the third angle. So, if two angles of one triangle are the same as two angles in another triangle, the third angle in both triangles will be the same. This is true even if the triangles are different sizes.

In this experiment you used triangulation by measuring the distance between two points (A and B), measuring the angles at A and B relative to the distant object, and using this information to draw a small triangle on a piece of graph paper. The drawn triangle is similar to (has the same angles as) the actual triangle you marked off with the three points (A, B, and the distant object). Because you used your feet to measure the actual distance between point A and point B, you know how far it is from point A to point B. By assigning a value of one foot to each square on the graph paper, you can then use the grid squares to estimate the distance between point A and the distant object by counting the number of squares between them on the graph paper.

In astronomy the technique of triangulation is called parallax. Centuries before the invention of the telescope, parallax was used by early astronomers to estimate distances between Earth and faraway planets and stars. Early astronomers could see a difference in the viewing angle of a star by looking at the star one day and then looking at it again months later. By measuring each of the two viewing angles, astronomers could calculate a distance to the faraway object. Using parallax measurements works well for stars that are closer than 400 light-years away. A light-year is the unit of distance that light will travel in one year. Since light travels at a velocity of about 300,000 km per second (186,411 mi./sec.), in one light-year the distance light travels is almost 10 trillion km or about 63,240 AU (astronomical units) with 1 AU being about 150 million km (93 million mi.)—the distance from Earth to the Sun.

ASTRONOMY

V. Just For Fun

Part A. Find another distant object that you can easily walk to (for example, a pole or other object in the backyard, park, field, or city street). Choose an object that you can walk to in a straight line without needing to go around anything. Repeat the experiment. In the space below, record your data.

Part B. After you have made your graphed triangle, check your calculation by measuring the distance to the distant object by walking heel-to-toe and counting each step as one "foot." Record your results.

A. Calculated Distance to Object

Number of steps—Point A to Point B _____

Angle at Point B _____

Number of squares—Point A to distant object _____

Calculated distance to object in "feet" _____

B. Measured Distance to Object

Measured distance to object in "feet" _____

Are the results of your calculation and your actual measurement the same or different? Why do you think you got this result?

_____

_____

_____

_____

_____

ASTRONOMY

Experiment 19

Using a Star Map

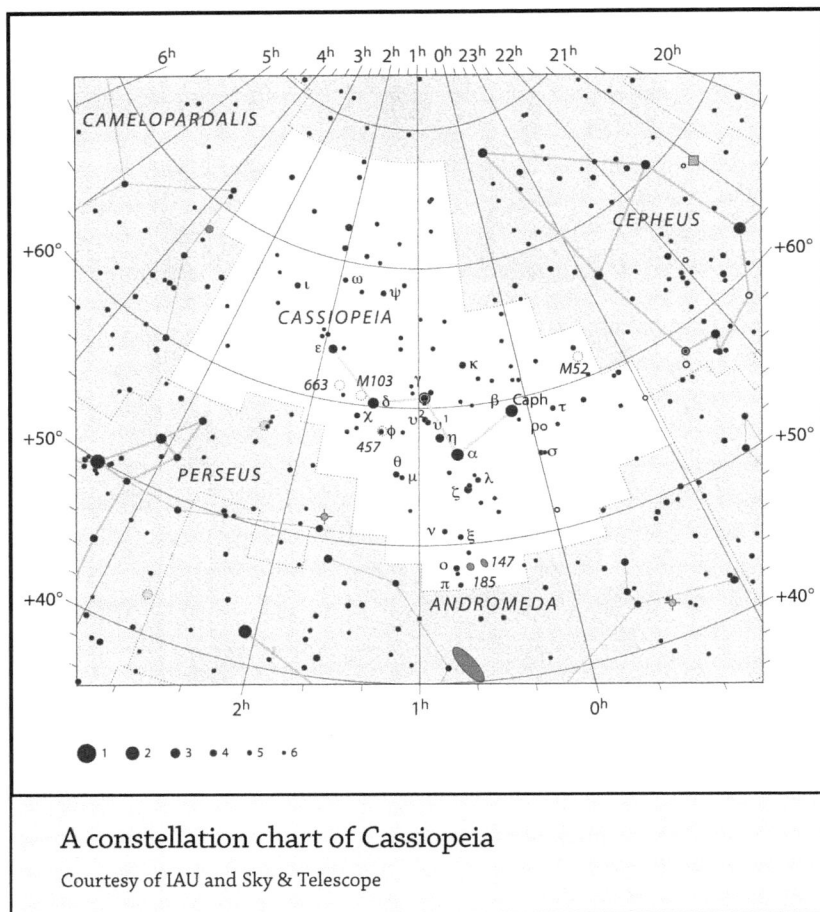

A constellation chart of Cassiopeia

Courtesy of IAU and Sky & Telescope

Introduction

Identify stars!

I. Think About It

❶ How many stars do you think are in the night sky? Why?

_____

_____

_____

_____

_____

❷ Do you think you could count all the stars you see? Why or why not?

_____

_____

_____

_____

_____

❸ Do you think you would see different stars in the night sky if you were at the North Pole than if you were at the South Pole? Why or why not?

_____

_____

_____

_____

ASTRONOMY

❹ If you had to map all the stars, how would you do it?

_____

_____

_____

_____

❺ If you could travel back in time to the first century, do you think your star map would look different? Why or why not?

_____

_____

_____

_____

❻ If you had an accurate map of the stars, in what ways might you use it?

_____

_____

_____

_____

ASTRONOMY

II. Experiment 19: Using a Star Map

Date _____

Objective

_____

_____

Materials

computer with internet access
printer and paper
flashlight

Optional

binoculars or telescope

EXPERIMENT

❶ Go to the Starmap site at http://www.star-map.fr/ and click on the Free Maps menu tab. This brings up a list of maps for the Northern Hemisphere. Under this list are three dots. Clicking on these will change the screen to show maps for the Equatorial Zone or the Southern Hemisphere.

Another free star map resource is http://www.skymaps.com.

[Note: Websites do change over time. If these sites are no longer available, do a browser search for "free star maps" to find a different resource.]

❷ On the Starmap site, select the hemisphere where you live, the map version you want to download, and the time you want to view the stars (20 pm is 8:00 pm and 22 pm is 10:00 pm). Click to download the map. Print the map.

❸ Study the star map you've downloaded. Check to make sure the map is for the correct hemisphere. Read the comments on the left side of the map. Many of the stars and constellations can be seen with an unaided eye, but some may require binoculars or a telescope.

❹ On an evening that is clear of clouds, go outside at the time you chose for the map you printed. Spend some time looking at the stars.

If you live in a city that has too much light at night for many stars to be seen, you may need to find a darker location away from city lights.

ASTRONOMY

❺ Hold the map above your head so you are looking up at it and the sky. See if you can orient the map to match the stars you observe in the sky. Note which stars you are able to identify.

❻ In the space provided in the *Results* section, use the star map as a guide to make your own star map by recording the constellations you saw and the magnitude of the stars. Add planets or other objects and their location.

Results

Star Map   Date_____

III. Conclusion

A. Questions

❶ Based on your observations how easy was it to locate the constellations? Why?

_____

_____

_____

_____

❷ How many constellations were you able to record? Which ones?

_____

_____

_____

_____

❸ Were there any constellations you were not able to find? Why?

_____

_____

_____

_____

❹ How does artificial lighting affect your viewing of the stars?

_____

_____

_____

_____

ASTRONOMY

❺ If you were to view the stars for several hours, do you think you would be able to observe whether the constellations move? Why or why not?

_____

_____

_____

_____

B. Conclusions

Based on your observations, what conclusions can you draw?

_____

_____

_____

_____

_____

_____

_____

_____

_____

_____

_____

ASTRONOMY

IV. Why?

By using a star map or star atlas you can become familiar with the stars. Like using a road map to find landmarks on Earth, you can learn landmarks in the night sky. If you were to study the night sky every night, you would become so familiar with it that you could navigate by the stars without using a star map.

There are many different kinds of star maps and star atlases available today. Some star maps show just the stars with the overlapping constellations. Some star maps have thousands of stars to observe and some only have a few hundred. There are also 3D star maps and planetary sphere maps that show star locations relative to your Earth position in a three-dimensional shape.

There are also deep space star maps that map the stars seen by the Hubble and Hipparcos telescopes. As astronomers continue to explore the cosmos, more stars and other celestial bodies are being added to star atlases.

V. Just For Fun

Repeat the experiment in a month. Download a new star map for that date and make your own star map based on the current location of the stars.

Compare your two star maps. What can you observe that is the same and what is different? The following pages provide space for your map and observations.

Star Map 2

Date _____

Star Map Comparisons

Star Map 1

Star Map 2

Experiment 20

Modeling Our Solar System

Introduction

Make a model of the planetary orbits of our solar system.

I. Think About It

❶ What do you think Earth would be like if it were in Mercury's orbit?

_____

_____

_____

_____

_____

❷ Do you think life as we know it could exist on Jupiter? Why or why not?

_____

_____

_____

_____

_____

❸ What do you think life on Earth would be like if it orbited two suns at the same time?

_____

_____

_____

_____

_____

ASTRONOMY

❹ What do you think it would be like if Earth's orbit was long and narrow instead of being almost round?

_____

_____

_____

_____

_____

❺ What do you think would happen if some planets orbited the Sun in a clockwise motion and others moved counterclockwise?

_____

_____

_____

_____

_____

❻ Why do you think planetary orbits are almost circular?

_____

_____

_____

_____

_____

ASTRONOMY

II. Experiment 20: Modeling Our Solar System Date _____

Objective _____

_____

Hypothesis _____

_____

_____

Materials

8 objects of different sizes to represent the planets
ruler (in centimeters)
marker
large flat surface for drawing—1 x 1 meter (3 x 3 feet), such as a
    large piece of cardboard or several sheets of construction paper
large open space at least 3 meters (10 feet) square
push pin
piece of string one meter (3 feet) long
tape

EXPERIMENT

❶ Find eight objects to represent the planets. Refer to the textbook illustration for the relative size of the planets and choose your objects to represent these sizes.

❷ Take the cardboard and mark the center with a marker. This represents the position of the Sun.

❸ Using the push pin, fix the string to the center mark of the cardboard.

❹ Measure 10 cm from the center and put a mark there. Wrap the loose end of the string around the marking pen so when the string is stretched out, the marking pen will be at the 10 cm mark. With the marker point touching the cardboard, draw a circle around the center mark. This is the orbital path for Earth.

☆☆ **ASTRONOMY**

❺ Draw concentric circles for the first 5 planetary orbits (Mercury through Jupiter) using the distances listed below. You will need to adjust the length of the string for each orbit.

❻ Place the objects you have chosen as your planetary models for the first 5 planets at their corresponding orbital distance from the center.

❼ For the last three orbits, measure the correct distance away from the center. Place the appropriate planetary model at the distance of its orbit.

Planet	Distance from Center
Mercury	4 cm
Venus	7 cm
Earth	10 cm
Mars	15 cm
Jupiter	50 cm
Saturn	90 cm (3 ft)
Uranus	190 cm (6 ft)
Neptune	300 cm (10 ft)

Results

Observe your model of the solar system and compare it with the illustration in your *Student Textbook*. On the following page, note any similarities or differences between your model of the solar system and the illustration. What else can you observe about the solar system?

Similarities	Differences
_____	_____
_____	_____
_____	_____
_____	_____
_____	_____
_____	_____
_____	_____
_____	_____
_____	_____
_____	_____
_____	_____
_____	_____
_____	_____
_____	_____
_____	_____
_____	_____

III. Conclusion

How easy or difficult was it to create a model of the solar system? How did the different distances affect how you could build your model? What did you learn by building the model?

_____

_____

_____

_____

_____

_____

_____

_____

_____

_____

_____

_____

_____

ASTRONOMY

IV. Why?

In this experiment you explored the orbital paths of the planets and how the planets are ordered in the solar system. An orbit is defined as the curved path that one celestial body follows as it travels around another celestial body. Although at one time it was thought that the Earth was the center of the universe and all the other celestial bodies orbited Earth, we now know that the Sun is the center of our solar system, making it a heliocentric system.

Basic physics tells us that bodies of mass have gravitational force, or gravity. The larger the body of mass, the more gravitational force it will have. The Sun is a very large body of mass and therefore has very strong gravitational force. Gravitational force keeps the planets in orbit around the Sun. The motion of the planets and the fact that the gravitational force of the Sun is constant are the things that keep the planetary orbits from collapsing towards the center of the solar system. The orbits of the planets are elliptical, but only slightly.

Mercury is in orbit closest to the Sun, and Neptune is farthest from the Sun. Measuring planetary distances is challenging because these distances are huge, and to show the distances in kilometers or miles results in very big numbers. To make it easier, astronomers use a unit of measure called the astronomical unit (AU) when talking about planetary distances. The distance from the Earth to the Sun is defined as 1 AU and the other planetary distances are some fraction or multiple of 1 AU. An AU is defined as the distance from Earth to the Sun because the distance of a planet from the Sun can be calculated using triangulation methods that require Earth's distance from the Sun as part of the calculation. Triangulation, or parallax, is still used as a method to arrive at distances and was used by ESA's Hipparcos satellite to accurately map the distances of over 100,000 stars. Radar and other methods are now also used to calculate distances.

The solar system can be divided into two different groups of planets according to their distance from the Sun. These groups are called the inner solar system and the outer solar system. There is a huge 4 AU gap between Mars (the outer planet of the inner solar system) and Jupiter (the inner planet of the outer solar system), and the Asteroid Belt is found in this gap.

ASTRONOMY

V. Just For Fun

Expand the features of your solar system model.

Find additional items to add the Asteroid Belt to your model. Would there also be asteroids outside the Asteroid Belt? Would you see comets somewhere? Would you see moons or any artificial satellites orbiting any planets? Would you see any space probes or landers? What would they be looking for? What else might you add to your solar system model?

Expanded Solar System Model

Discovering Life on Other Planets

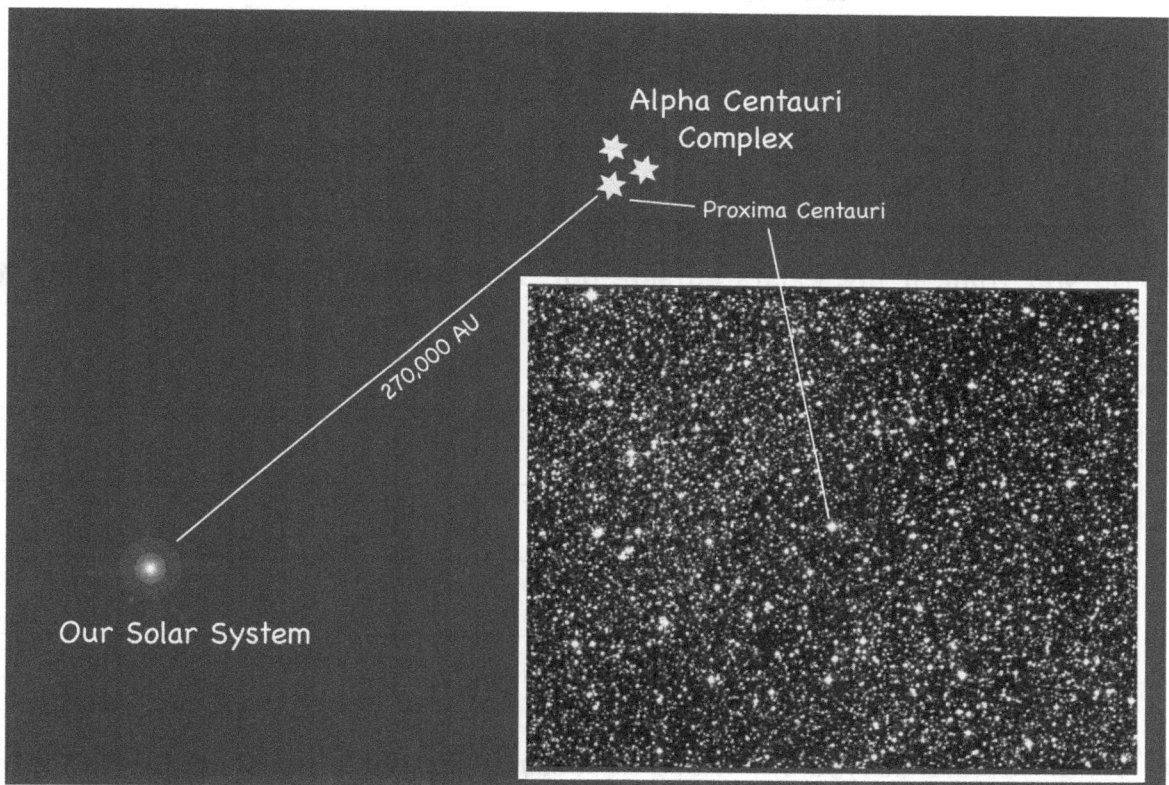

Introduction

Explore thought experiments.

I. Think About It

❶ What do you think is the likelihood that there is life on planets and/or moons outside our solar system? Why?

_____

_____

_____

_____

_____

❷ Do you think if life exists outside our solar system, it would be similar to life on Earth? Why or why not?

_____

_____

_____

_____

_____

❸ What factors do you think are necessary for life to exist on Earth? Why?

_____

_____

_____

_____

_____

ASTRONOMY

❹ Do you think life outside our solar system would require the same conditions that we have on Earth? Why or why not?

_____

_____

_____

_____

_____

❺ How do you think you would detect "life" on another planet?

_____

_____

_____

_____

_____

❻ How do you think living on a moon would be different from living on a planet? Why?

_____

_____

_____

_____

_____

ASTRONOMY

II. Experiment 21: Discovering Life on Other Planets

Date _____

Thought Experiment

Sometimes when it's not possible to do an actual experiment, it can be very useful to do what is called a *thought experiment.* A thought experiment is a mental exercise in which an experiment is imagined. The process of imagining how a hypothesis might be explored or how an experiment might actually work is very valuable to science. Albert Einstein wondered what it would be like to ride on a rainbow. He could not literally ride on a rainbow, but he could imagine it, and the ideas he generated during this thought experiment helped him discover the theory of relativity.

Materials

pencil
colored pencils
your imagination

EXPERIMENT

❶ Imagine that you are traveling outside our solar system and you come across a star three times the size of our Sun. You observe ten planets in the solar system around this sun. Some of the planets have moons. Assume that you can travel to all ten planets and explore all of their moons.

❷ Do a thought experiment and write in as much detail as possible what you would need to do to locate life on any of the ten planets or moons. Imagine this is really possible. Think about what you would need to take with you and how you would define "life." Also consider which planets or moons are more likely to have life and which you can ignore.

ASTRONOMY

Discovering Life—A Thought Experiment

_____

_____

_____

_____

_____

_____

_____

_____

_____

_____

_____

_____

_____

_____

_____

_____

_____

_____

_____

_____

ASTRONOMY

ASTRONOMY

III. Why?

Because the universe is home to billions of stars, it makes sense to assume that some of those stars have planets. The idea of the existence of other planets has fascinated both scientists and science fiction writers for many years, but the existence of exoplanets has only recently been confirmed.

The idea that there might be life on other planets comes from the overwhelming number of possible planets that could orbit the billions of stars in the universe. Because we know the criteria for life on Earth, astronomers can begin to look at stars and the planets that orbit those stars to determine if there are any planets that meet the criteria for life as we know it.

Finding and studying exoplanets is extremely challenging. Most exoplanets lie close to their parent star. Direct imaging is difficult because the light from the star hides the planets. However, an exoplanet can be observed by direct imaging if the parent star is weakly luminous or if the exoplanet has a wide orbit.

Exoplanets can be indirectly observed by analyzing light from the parent stars. Recall that planets have mass and because of this have gravitational force. When a planet is orbiting a star, the star may "wobble" as a result of the planet's gravitational pull. The more massive the planet and the less massive the star, the more the star wobbles. Astronomers can use the wobbling of a star to estimate the mass of its exoplanets. Also, as a planet passes in front of the star it is orbiting, the light from the star dims slightly as the planet blocks some of the light. This slight dimming can be detected and used as another indirect method of finding exoplanets.

Although many exoplanets have been discovered, scientists are just beginning to find planets that might have the right conditions to be suitable to support life as we know it. It is thought that in order to support life, an exoplanet must be just the right distance from the parent star—neither too close nor too far away. This "Goldilocks distance" is called the Circumstellar Habitable Zone.

As technology advances, more and more discoveries will be made about exoplanets and the possibility that certain exoplanets could support life. The hope is that one day technology will advance enough that we can travel to distant planets to look for life and discover more about the universe we live in.

ASTRONOMY

IV. Just For Fun

Review your thought experiment notes. In the space below make a diagram of the solar system you've explored, including the sun and the planets and any moons. Draw the orbits of the planets and name the solar system and the different celestial bodies in your diagram.

Diagram of the _____ **Solar System**

ASTRONOMY

Choose one planet or moon where you imagine you have found life. Draw and/ or write about the different life forms you found. How did you find them? What do they look like? What are they doing? What conditions do they need to live? (More space on next page.)

Life on (Planet or Moon) _____

Life Information (Continued)

ASTRONOMY

Experiment 22

Working Together

Introduction

Explore working on a collaborative science project.

I. Think About It

❶ Do you think it is helpful for scientists who live in different countries to work together on a particular research project? Why or why not?

_____

_____

_____

_____

❷ How many kids do you know who do real science experiments? What kinds of experiments do you think they do?

_____

_____

_____

_____

❸ How easy or difficult do you think it would be for you to collaborate on a science experiment with other kids in your town?

_____

_____

_____

_____

❹ If you were to collaborate with kids doing science in another country, how would you do it?

_____

_____

_____

_____

_____

❺ Do you think you could learn more by doing science with other kids as collaborators than by doing experiments on your own? Why or why not?

_____

_____

_____

_____

_____

❻ Do you think some kinds of experiments might work well with collaborators and others kinds of experiments might not? Why?

_____

_____

_____

_____

_____

II. Experiment 22: Working Together   Date   _____

Objective   _____

_____

Hypothesis   _____

_____

_____

Materials

computer with internet access
any materials required for the experiment you choose

EXPERIMENT

❶ Go to the SciStarter website:
http://scistarter.com/index.html

❷ Click on "Pick an Activity" or "Pick a Topic."

❸ Use the drop-down menu to choose a category, then review some of the experiments that will come up. A few possibilities for projects:

- Measure the Vitamin C in your food (Topic--> Food --> *Measuring the Vitamin C in Food*).

- Grow crystals (Topic--> Chemistry---> *The Art of Crystallisation - a global experiment*).

- Observe dogs' favorite poo places (*Poo Power! Global Challenge*—do a search on the website for Poo Power).

- Collect and send soil samples (Activity--> At Home--> *Drug discovery from your soil*).

- Identify clouds (Activity--> On a hike--> *Identify the Cloud*).

- Pick a *Project of the Day.*

Not all projects will fit your schedule and location, so you may need to do a bit of searching.

In the following space, write down the names of experiments that look interesting to you.

_____

_____

_____

_____

_____

_____

❹ From the above list, select one activity you would like to do. To find a project again, you can do a search on the name you've written down.

❺ Click on the *Get Started* button and follow the instructions. Gather any materials that may be required.

❻ Perform your experiment or observations and follow the directions for submitting your results.

Results

On the following page, record the name of the project you participated in and the name of the organization that is sponsoring it. Describe what you did, what results you got, and how you think the data will be used. Record any additional observations.

Project: _____

Sponsor: _____

Description of the experimental process, results, how the data will be used, and other observations

III. Conclusions

Discuss what you learned by participating in a crowdsourced scientific experiment.

_____

_____

_____

_____

_____

_____

_____

_____

_____

_____

_____

_____

_____

_____

_____

_____

_____

IV. Why?

Collaborating on projects is a wonderful way to help solve real problems and learn how other scientists investigate a variety of projects. Although science continues to be a solitary career for many scientists working in research labs, being able to share data and participate in large collaborative projects has the potential to accelerate discoveries, enriching the lives of both scientists and non-scientists alike.

In this experiment you participated in a collaborative project as a citizen scientist. Although you don't yet have a degree in science or work for a large lab or research company, you can still participate in collecting data and observing how your data gets used to solve real problems. You can learn more about what kinds of projects scientists study, how they set up experiments, and how they collect and report data.

V. Just For Fun

Make up your own collaborative experiment using the format of the other experiments in this book. Include a title, an objective and a hypothesis, a materials list, a list of those who will be participating, steps to be followed, results, and conclusions. Also include any charts or drawings that are needed. Choose from the suggestions below or use your own idea.

- **Chemistry: What will make an acid-base indicator?** Have your collaborators join you in finding fruits and vegetables that you can test to see if they can be used as acid-base indicators. Refer to the *Just For Fun* section in Experiment 3. Find plants that you haven't already tested—for example, mango, pumpkin, rhubarb.

- **Biology: Find out what's in dirt!** Have each participant dig up a small sample of dirt from their yard or somewhere else where it's OK for them to dig. Have each collaborator examine their dirt sample under a microscope and record their observations in as much detail as possible. Put the results in a chart and compare them. What is the same and what is different about the samples? Did anyone see something they found surprising? Have a discussion about the results.

- **Physics: Be a coach!** Have your collaborators become a running or cycling (or other) team. Use Experiment 12 as a guide for doing the experiment and collecting and analyzing data for each participant. Combine and analyze the data at the end of the experiment. What did you find out?

- **Geology: What's in your neighborhood?** Think about what you would like to learn about your neighborhood and join with your collaborators in making a map. For example, each participant can take one area of your neighborhood and make a piece of a map that can be assembled at the end of the experiment. Or each participant can map a different feature of the same area (such as streets, houses, plants, animals, geological formations, etc.) and then each can draw their discoveries on the same map.

- **Astronomy: Think about Mars!** Join your collaborators in doing a thought experiment about what it would take to start a colony on Mars and what the colony would look like. Each participant can do the thought experiment on their own, writing down their ideas, or the group can discuss ideas with someone recording what is said. Think about how you would categorize the ideas and then record all the ideas in an organized chart. Have the collaborators make drawings and models based on these ideas.

More REAL SCIENCE-4-KIDS Books
by Rebecca W. Keller, PhD

Building Blocks Series
yearlong study program — each Student Textbook has accompanying Laboratory Notebook, Teacher's Manual, Lesson Plan, Study Notebook, Quizzes, and Graphics Package

Exploring Science Book K (Activity Book)
Exploring Science Book 1
Exploring Science Book 2
Exploring Science Book 3
Exploring Science Book 4
Exploring Science Book 5
Exploring Science Book 6
Exploring Science Book 7
Exploring Science Book 8

Focus On Series
unit study program — each title has a Student Textbook with accompanying Laboratory Notebook, Teacher's Manual, Lesson Plan, Study Notebook, Quizzes, and Graphics Package

Focus On Elementary Chemistry
Focus On Elementary Biology
Focus On Elementary Physics
Focus On Elementary Geology
Focus On Elementary Astronomy

Focus On Middle School Chemistry
Focus On Middle School Biology
Focus On Middle School Physics
Focus On Middle School Geology
Focus On Middle School Astronomy

Focus On High School Chemistry

Super Simple Science Experiments

21 Super Simple Chemistry Experiments
21 Super Simple Biology Experiments
21 Super Simple Physics Experiments
21 Super Simple Geology Experiments
21 Super Simple Astronomy Experiments
101 Super Simple Science Experiments

Note: A few titles may still be in production.

Gravitas Publications Inc.
www.gravitaspublications.com
www.realscience4kids.com

GRAVITAS
PUBLICATIONS